SpringerBriefs in Population Studies

Population Studies of Japan

The world population is expected to expand by 39.4% to 9.6 billion in 2060 (UN World Population Prospects, revised 2010). Meanwhile, Japan is expected to see its population contract by nearly one third to 86.7 million, and its proportion of the elderly (65 years of age and over) will account for no less than 39.9% (National Institute of Population and Social Security Research in Japan, Population Projections for Japan 2012). Japan has entered the post-demographic transitional phase and will be the fastest-shrinking country in the world, followed by former Eastern bloc nations, leading other Asian countries that are experiencing drastic changes.

A declining population that is rapidly aging impacts a country's economic growth, labor market, pensions, taxation, health care, and housing. The social structure and geographical distribution in the country will drastically change, and short-term as well as long-term solutions for economic and social consequences of this trend will be required.

This series aims to draw attention to Japan's entering the post-demographic transition phase and to present cutting-edge research in Japanese population studies. It will include compact monographs under the editorial supervision of the Population Association of Japan (PAJ).

The PAJ was established in 1948 and organizes researchers with a wide range of interests in population studies of Japan. The major fields are (1) population structure and aging; (2) migration, urbanization, and distribution; (3) fertility; (4) mortality and morbidity; (5) nuptiality, family, and households; (6) labor force and unemployment; (7) population projection and population policy (including family planning); and (8) historical demography. Since 1978, the PAJ has been publishing the academic journal *Jinkogaku Kenkyu* (The Journal of Population Studies), in which most of the articles are written in Japanese.

Thus, the scope of this series spans the entire field of population issues in Japan, impacts on socioeconomic change, and implications for policy measures. It includes population aging, fertility and family formation, household structures, population health, mortality, human geography and regional population, and comparative studies with other countries.

This series will be of great interest to a wide range of researchers in other countries confronting a post-demographic transition stage, demographers, population geographers, sociologists, economists, political scientists, health researchers, and practitioners across a broad spectrum of social sciences.

More information about this subseries at http://www.springer.com/series/13101

Fumiya Uchikoshi · James M. Raymo

Educational Assortative Mating in Japan

Insights into Social Change and Stratification

Springer

Fumiya Uchikoshi
Department of Sociology and Office
of Population Research
Princeton University
Princeton, NJ, USA

James M. Raymo
Department of Sociology and Office
of Population Research
Princeton University
Princeton, NJ, USA

ISSN 2211-3215 ISSN 2211-3223 (electronic)
SpringerBriefs in Population Studies
ISSN 2198-2724 ISSN 2198-2732 (electronic)
Population Studies of Japan
ISBN 978-981-16-3712-4 ISBN 978-981-16-3713-1 (eBook)
https://doi.org/10.1007/978-981-16-3713-1

This Springer imprint is published by the registered company Springer Nature Singapore Pte Ltd.
The registered company address is: 152 Beach Road, #21-01/04 Gateway East, Singapore 189721, Singapore

Preface

In the Economist Aug 20th 2011 issue, the headline article titled "Asia's lonely hearts" described how and why Asian women, especially highly educated ones, increasingly avoid marriage. The article argued that women's rejection of marriage is a result of strong societal expectations that women devote themselves to their families even if they are employed full-time. This same argument has been made in Japan for a long time; demographers consistently concluded that later and less marriage was driven primarily by women's economic independence and the incompatibility of their role as caregiver after marriage and their opportunities in the labor market (Raymo 2003).

The Economist article goes further, however, in pointing to the social norm that encourages women to "marry up." In the social sciences, this is called female hypergamy, or simply hypergamy, with the focus on women implied. This hypergamy norm was rooted in a gender-inegalitarian social environment in which women's socio-economic status was systematically lower than men's. However, as women's educational attainment has become increasingly comparable to, and even surpassed, that of men, highly educated women in gender-inegalitarian societies encounter increasing numeric difficulty in finding a suitably educated partner. This strong preference for women's educational hypergamy is seen as another crucial factor in explanations of declining marriage rates in East Asian countries (Ji and Yeung 2014; Mu and Xie 2014; Raymo and Iwasawa 2005; Raymo and Park 2020).

As such, the persistent gender inequality characterizing Japanese society (Brinton 1993; Brinton and Oh 2019) has reinforced women's preferences for status hypergamy and men's preferences for marrying a spouse who will focus primarily on providing household labor (Brinton et al. 2021). However, there is a good reason to believe that this conventional wisdom may be in need of an update, as men's economic standing gradually declines (Brinton 2011), and both men and women increasingly prefer dual-earner households (IPSS 2017). In this emerging context, some studies have reported an increase in women's educational hypogamy (marrying down) (Fukuda et al. 2020). If views of hypogamy as less desirable or culturally deviant resulted in such marriages being at higher risk of dissolution in the past, the trend toward more educational hypogamy suggests such gradients in marital stability may have disappeared in recent years (Schwartz and Han 2014; Uchikoshi 2019).

Changes in spouse pairing patterns, or patterns of assortative mating, are an important feature of family changes taking place in many low fertility countries. There is abundant evidence of an increase in educational homogamy (marrying similarly educated partners),[1] especially at the top of educational distribution, in several countries as a result of increasing economic sorting (Schwartz and Mare 2005), with potential implications for economic inequality (Schwartz 2010). Changing spouse pairing patterns may also have implications for the reproduction of (dis)advantage, and social mobility more generally, to the extent that growth in homogamy at both ends of the educational distribution exacerbates the trend of diverging life course outcomes by educational attainment (Lundberg et al. 2016; McLanahan 2004; Reeves 2017). Importantly, the same concern has been expressed by scholars working on Japan (Raymo and Iwasawa 2016; Tachibanaki and Sakoda 2013; Tsutsui 2016).

Echoing these studies, we believe that there is value in examining not only trends in patterns of educational assortative mating, but also their consequences. While there have been some efforts to examine these questions in Western contexts, there is no study (to the best of our knowledge) that evaluates the degree to which assortative mating matters for family and inequality in non-Western contexts. In this manuscript, we tackle this question by focusing on Japan, a society characterized by both profound change and continuity in social norms and expectations that may have broader relevance across East Asia (Raymo et al. 2015).

Princeton, USA

Fumiya Uchikoshi
James M. Raymo

References

Brinton, Mary C. 1993. *Women and the economic miracle: Gender and work in postwar Japan*. California: University of California Press.

Brinton, Mary C. 2011. *Lost in Transition: Youth, Work, and Instability in Postindustrial Japan*. Cambridge: Cambridge University Press.

Brinton, Mary C., and Eunsil Oh. 2019. Babies, work, or both? Highly educated women's employment and fertility in East Asia. *American Journal of Sociology* 125 (1): 105–140. https://doi.org/10.1086/704369.

Brinton, Mary C., Eunmi Mun, and Ekaterina Hertog. 2021. Singlehood in contemporary Japan: Rating, dating, and waiting for a good match. *Demographic Research* 44: 239–276. https://doi.org/10.4054/DemRes.2021.44.10.

Fukuda, Setsuya, James M. Raymo, and Shohei Yoda. 2020. Revisiting the educational gradient in marriage in Japan. *Journal of Marriage and Family* 82 (4): 1378–1396. https://doi.org/10.1111/jomf.12648.

[1] Throughout the text, we primarily use "assortative mating" to refer to pairing patterns in general while "homogamy" is used to refer to a specific pairing pattern, marrying someone with same characteristics in this case. However, since the dominant spouse pairing pattern is homogamy, we occasionally use the term assortative mating to refer generally to homogamy. For example, the statement that assortative mating may be a driving force of income inequality assumes that most marriages are homogamous.

Ji, Yingchun, and Wei-Jun Jean Yeung. 2014. Heterogeneity in contemporary Chinese marriage. *Journal of Family Issues* 35 (12): 1662–1682. https://doi.org/10.1177/0192513X14538030.

Lundberg, Shelly, Robert A. Pollak, and Jenna Stearns. 2016. Family inequality: Diverging patterns in marriage, cohabitation, and childbearing. *Journal of Economic Perspectives* 30 (2): 79–102. https://doi.org/10.1257/jep.30.2.79.

McLanahan, Sara. 2004. "Diverging destinies: How children are faring under the second demographic transition. *Demography* 41 (4): 607–627. https://doi.org/10.1353/dem.2004.0033.

Mu, Zheng, and Yu Xie. 2014. Marital age homogamy in China: A reversal of trend in the reform era? *Social Science Research* 44: 141–157. https://doi.org/10.1016/j.ssresearch.2013.11.005.

Raymo, James M. 2003. Educational attainment and the transition to first marriage among Japanese women. *Demography* 40 (1): 83–103. https://doi.org/10.1353/dem.2003.0008.

Raymo, James M., and Hyunjoon Park. 2020. Marriage decline in Korea: Changing composition of the domestic marriage market and growth in international marriage. *Demography* 57 (1): 171–194. https://doi.org/10.1007/s13524-019-00844-9.

Raymo, James M., and Miho Iwasawa. 2005. Marriage market mismatches in Japan: An alternative view of the relationship between women's education and marriage. *American Sociological Review* 70 (5): 801–822. https://doi.org/10.1177/000312240507000504.

Raymo, James M., and Miho Iwasawa. 2016. *Diverging destinies: The Japanese case*. Springer.

Raymo, James M., Hyunjoon Park, Yu Xie, and Wei-Jun Jean Yeung. 2015. Marriage and family in East Asia: Continuity and change. *Annual Review of Sociology* 41: 471–92. https://doi.org/10.1146/annurev-soc-073014-112428.

Reeves, Richard V. 2017. *Dream hoarders: how the American upper middle class is leaving everyone else in the dust, why that is a problem, and what to do about it*. Washington, D.C.: Brookings Institution Press.

Schwartz, Christine R. 2010a. Earnings inequality and the changing association between spouses' earnings. *American Journal of Sociology* 115 (5): 1524–1557. https://doi.org/10.1086/651373.

Schwartz, Christine R. 2010b. Pathways to educational homogamy in marital and cohabiting unions. *Demography* 47 (3): 735–753.

Schwartz, Christine R., and Hongyun Han. 2014. The reversal of the gender gap in education and trends in marital dissolution. *American Sociological Review* 79 (4): 605–629. https://doi.org/10.1177/0003122414539682.

Schwartz, Christine R., and Robert D. Mare. 2005. Trends in educational assortative marriage from 1940 to 2003. *Demography* 42 (4): 621–646.

Tachibanaki, Toshiaki, and Sayaka Sakoda. 2013. *Fufu Kakusa Shakai (Couples Disparity Society)*. Chuo Koronsha (in Japanese).

Tsutsui, Junya. 2016. *Kekkon to Kazoku No Korekawa (The Future of Marriage and Family)*. Kobunsha. (in Japanese).

Uchikoshi, Fumiya. 2019. Consequences of educational assortative mating through divorce. *Japanese Sociological Review* 70 (1): 10–26 (in Japanese).

Acknowledgments

This research is supported by Nakajima Foundation and JSPS Grant-in-Aid for Specially Promoted Research (Grant number 25000001 and the Eunice Kennedy Shriver National Institute of Child Health & Human Development of the National Institutes of Health under Award Number P2CHD047879). We thank the 2015 SSM Survey Management Committee for allowing Fumiya Uchikoshi to use the SSM data. The data for this secondary analysis, Japanese Life Course Panel Survey of the Youth (JLPS-Y) and the Middle-aged (JLPS-M) Waves 1–9, 2007–2015 (Japanese Life Course Panel Surveys project, Institute of Social Science, The University of Tokyo), were provided by the Social Science Japan Data Archive, Center for Social Research and Data Archives, Institute of Social Science, The University of Tokyo.

Contents

About the Authors

Fumiya Uchikoshi is a Ph.D. student in Sociology at Princeton University. His research interests include family demography, social stratification, and East Asia. His current research examines diverging family behaviors and their impact on social inequality and the consequences of newly emerging behaviors on future life course outcomes in familistic societies.

James M. Raymo is Professor of Sociology and the Henry Wendt III '55 Professor of East Asian Studies at Princeton University. Raymo is a social demographer whose research focuses on documenting and understanding the causes and potential consequences of demographic changes in Japan. His published research includes analyses of marriage timing, divorce, recession and fertility, marriage and women's health, single mothers' well-being, living alone, employment and health at older ages, and regional differences in health at older ages. His current research focuses on children's well-being, changing patterns of family formation, single motherhood, and social isolation and health at older ages.

About the Authors

Abbreviations

IPSS	National Institute of Population and Social Security Research
JLPS	Japanese Life-course Panel Study
MEXT	Ministry of Education, Culture, Sports, Science and Technology
NFS	National Fertility Survey
SSM	Social Stratification and Mobility Survey

List of Figures

List of Tables

Chapter 1
Introduction

In this introductory chapter, we begin by briefly reviewing why we need to care about assortative mating. We focus on three specific, substantive topics in assortative mating research. First, the question of who marries whom is of great interest to demographers and stratification researchers who are interested in how spouse pairing patterns reflect existing societal boundaries and openness. Second, research on assortative mating suggests that an increase in educational homogamy at the top and bottom of the educational distribution may have implications for economic inequality and intergenerational transmission of (dis)advantage. Third, recent studies have shown that changing spouse pairing patterns have implications for our understanding of emerging trends in fertility, marital stability, and the gender division of labor. After reviewing these theoretical motivations for the study of assortative mating, we discuss the aims of this book as well as the rationale for focusing on Japan. We conclude this chapter by introducing the outline of this book.

1.1 Why Assortative Mating is Important

1.1.1 Assortative Mating as an Indicator of Societal Openness

A classic focus in research on spouse pairing is on assortative mating by socioeconomic status (SES) (e.g., class, income, and education) as a reflection of social boundaries between groups. A society is regarded as relatively closed if marriage patterns are predominantly homogamous (or endogamous). Attention to societal openness/closure is important given that the rigidity of boundaries associated with markers of status reflects structural factors that constrain opportunities for securing socioeconomic rewards (Collins 1979; Murphy 1984; Parkin 1971). While mate selection in the past was based primarily on ascribed status markers such as class, caste, religion, or inherited traits, assortative mating is increasingly based on achieved characteristics (Kalmijn 1991; Rosenfeld 2008), especially educational attainment.

F. Uchikoshi and J. M. Raymo, *Educational Assortative Mating in Japan*,
Population Studies of Japan, https://doi.org/10.1007/978-981-16-3713-1_1

This is particularly true in economically developed countries, where in the process of modernization individuals have increasingly come to value and pursue markers of achieved status (Treiman 1970). The centrality of educational attainment as a dimension of assortative mating has led scholars to examine the rigidity of educational boundaries in the marriage market and, by extension, in society more generally (Blossfeld 2009). With reference to modernization theory, scholars have hypothesized that the educational expansion accompanying modernization should result in an increase in educational homogamy (Smits et al. 1998), but empirical evidence is mixed (as we discuss below).

1.1.2 Assortative Mating as a Mechanism to Explain Socioeconomic Inequality

Increasing economic inequality in many rich countries since the mid-1970s (Brandolini and Smeeding 2009; Piketty 2014) has a variety of implications for demographic and intergenerational processes (Chetty et al. 2016; Corak 2013; Duncan and Murnane. 2011; Seltzer 2019). While many explanations for rising economic inequality focus on growing wage inequality by educational attainment or changing demands for skills in the labor market (Acemoglu and Autor 2011; Autor 2010; Goldin and Katz 2009), others focus on the role of changes in family formation, including educational assortative mating (Esping-Andersen 2007; McCall and Percheski 2010; McLanahan and Percheski 2008). This section reviews several studies that have provided important insights into how we understand the potential impact of changes in educational assortative mating on trends in inequality and socioeconomic mobility.

1.1.2.1 Income Inequality

According to McCall and Percheski (2010), the causes of increasing inequality can be divided into three sub-mechanisms: changes in family structure, women's improved access to the labor market, and an (absolute) increase in educational assortative mating. The growing prevalence of "fragile families," especially single-mother families, has been linked to increasing rates of poverty and inequality (McLanahan and Percheski 2008), especially in countries like the United States where public income support is limited (Gornick and Jäntti 2012). Evidence that married women's labor force participation has been negatively correlated with husbands' income suggests that women's increasing employment trends should contribute to a reduction in household income inequality (Cancian and Reed 1998; Kollmeyer 2013; Treas 1987), but recent evidence suggests a weakening of this negative correlation, at least in the United States (Blau and Kahn 2007; Killewald and Zhuo 2019; Shafer 2011).

It makes intuitive sense that an increase in educational homogamy, especially at the top and bottom of the educational distribution, should contribute to higher levels of inequality, but the implications of trends in educational assortative mating are the subject of much study and debate (Breen and Andersen 2012; Breen and Salazar 2010; Greenwood et al. 2014; Hu and Qian 2015; Monaghan 2015; Schwartz 2010a; Torche 2010; Western et al. 2008; Zagel and Breen 2019). Some recent studies find a null association between trends in educational assortative mating and growing household inequality in Britain (Breen and Salazar 2010), the United States (Greenwood et al. 2015; Hryshko et al. 2017), and four Western countries (Boertien and Permanyer 2019) including the United States (Eika et al. 2019). Others conclude that the rise in educational assortative mating contributed to a reduction in economic inequality in China (Hu and Qian 2015) and the United States (Breen and Salazar 2011). In societies characterized by a marked rise in women's labor force participation such as Denmark, research shows that the rise in educational assortative mating contributes to increasing income inequality (Breen and Andersen 2012; Esping-Andersen 2007), reflecting the facts that highly educated women are more likely to have full-time employment and that these highly educated women tend to marry a spouse with similar education (but see Boertien and Permanyer 2019 for a contrasting argument).

These varied conclusions suggest that the impact of educational assortative mating on economic inequality depends on the degree to which women's education is associated with employment and earnings (Schwartz 2013; Herzberg-Druker and Stier 2019). In this context, the life-course perspective (Gonalons-Pons and Schwartz 2017) provides critical insights into the relationship between educational assortative mating and income inequality, by focusing on how women's employment changes in response to life-course transitions. Prior studies have found that women's earning trajectories are more variable over the life course than men's and that earnings trajectories are influenced by spouse pairing patterns (Qian 2018). Specifically, women who marry down (educational hypogamy) tend to increase their earnings following marriage to a greater degree than women in homogamous or hypergamous marriages, suggesting that these women are more likely to have non-traditional gender attitudes and/or have greater bargaining power within marriage (Qian 2018: 618). A key insight from this research is that it is change in the gender division of labor after marriage, rather than changing patterns of sorting in the marriage market, that plays an important role in explaining the recent rise in economic homogamy in the United States (Gonalons-Pons and Schwartz 2017). In conjunction with evidence that women's educational hypogamy is less stigmatized in recent cohorts (Schwartz and Han 2014; Uchikoshi 2019), we can thus speculate that an increase in educationally hypogamous marriages characterized by high levels of wives' employment after marriage and childbearing could actually contribute to an increased correlation in spouses' earnings and aggregate economic inequality.

1.1.2.2 Intergenerational Mobility

Most previous studies on assortative mating focus on pairing patterns within relatively short periods of time (single generations), but emphasize assortative mating as an indicator of societal openness that provides a compelling link to research on intergenerational mobility. Despite the overlapping focus on the association between the socioeconomic status of two actors (husbands and wives in assortative mating research and parents and children in intergenerational mobility research), we are aware of only a few efforts to explicitly link the two (Mare 2016; Uchikoshi 2016).

There are at least two compelling reasons to pursue this linkage. First, studies on homophily (Kalmijn 1998) suggest that children raised by two similarly, highly educated parents would be more likely to interact with others having similar educational characteristics than would children of less-educated parents. Second, these children are expected, through processes of socialization, to possess cultural values and lifestyles that resemble those of their parents and their peers from similar family backgrounds (Kalmijn 1998: 418–419). Combining these two perspectives, we can hypothesize that educational assortative mating in the parents' generation transmits to that of children's generation (Mare 2016). Indeed, Mare's (2016) analysis of U.S. data showed that educational resemblance among parents is positively associated with the odds of their children marrying educationally similar spouses. Similar findings were reported by Uchikoshi (2016). Also, several studies have found that children with similar parental wealth are more likely to marry each other (Charles et al. 2013; Wagner et al. 2020), even after controlling for children's own educational attainment. Together, these findings suggest that parental characteristics (pairing patterns and wealth) shape assortative mating among their children in ways that may limit socioeconomic mobility across generations.

Economic models of family formation also point to ways in which patterns of assortative mating likely matter for intergenerational inequality. According to Lundberg et al.'s (2016) theoretical depiction of marriage as a commitment device, growth in cohabitation as a family alternative to marriage may change the cost–benefit calculations of marriage among young adults. Because the exit cost (of dissolution) is higher for marriage than for cohabitation, marriage may be increasingly characterized by stronger commitments, including investment in children's education. While it makes sense that married couples may invest more in children's education and while we know that both quantity and quality of children's education are strongly associated with well-being (broadly defined) across the life course, research on the potential pathways through which parents' educational assortative mating may shape the transmission of (dis)advantages to the children's generation is limited. Some studies have concluded that parental educational homogamy has a positive impact on children's health, perhaps reflecting lower levels of mothers' prenatal stress in such couples (Pesando 2021; Rauscher 2020). Independent effects of assortative mating (especially among those who are highly educated) on a child's cognitive outcomes are also reported in other studies (Bai 2018; Beck and González-Sancho 2009; Byun et al. 2020; Edwards and Roff 2016).

1.1.3 Impacts of Assortative Mating on Family Life

The changing gender landscape is of particular importance for understanding links between patterns of assortative mating and family formation behaviors (Esping-Andersen and Billari 2015; Goldscheider et al. 2015; van Bavel et al. 2018). With women now outnumbering men in higher education in most rich countries (DiPrete and Buchmann 2014; Esteve et al. 2012, 2016; van Bavel 2012), it is increasingly critical to ask how the reversal of the gender gap in education changes basic features of our society, including gender relations and marriage. In this context, the rise of educational hypogamy (women marrying down) provides important insights into how changing spouse pairing patterns shape family demographic outcomes, including fertility (Nitsche et al. 2018; Nomes and van Bavel 2016; Osiewalska 2017; Trimarchi and van Bavel 2020), divorce (Schwartz and Han 2014; Uchikoshi 2019), and the gender division of labor (Miller 2020).

1.1.3.1 Fertility

Relationships between women's educational attainment and fertility outcomes are well documented, but focusing on patterns of educational assortative mating allows us to also understand the influence of the male partner's education (i.e., the interactive association of spouses' educational attainment with fertility outcomes) (Dribe and Stanfors 2010). Conventional neoclassical economic theories of marriage emphasizing the pooling of complementary specializations (Becker 1991) suggest that the fertility of educationally hypergamous couples should be higher than that of other types of couples. Alternatively, evidence that educationally homogamous couples have similar attitudes and values (Kalmijn 1991) and are less likely to divorce (Schwartz 2010b; Tzeng 1992) suggests that the fertility of these couples may be higher than that of educationally heterogamous couples.

Reversal of the gender gap in higher education has motivated attention to the consequences of increasing educational hypogamy for fertility (Nomes and van Bavel 2016; Osiewalska 2017; van Bavel and Klesment 2017). Potential pathways through which educational hypogamy may affect fertility are: (1) higher opportunity costs of childbearing among households in which women outearn their husbands (van Bavel and Klesment 2017), (2) the difference in marital timing between lower- and higher-educated women, contributing to women in hypogamous unions postponing their childbearing, and (3) a positive correlation between women's willingness to marry down and preferences for fewer, or no, children (Nomes and van Bavel 2016). Interestingly, one study observed that rates of second birth are lowest among educationally hypergamous couples while rates of second and third births are highest among educationally homogamous couples (Nitsche et al. 2018). This study provides indirect evidence that hypotheses derived from the theoretical models of marriage associated with New Home Economics paradigm are no longer empirically supported, while another study suggests that the relationship between educational assortative mating and fertility varies by country context (Trimarchi and van Bavel 2020).

1.1.3.2 Divorce

Assortative mating is also critical for understanding relationship stability. Scholars have long suggested that educationally hypogamous couples tend to be culturally dissimilar from one another (Kalmijn 1991) and to deviate from prevailing social norms that used to favor male breadwinner-female homemaker marriages and now favor egalitarian dual-earner couples. Among couples in which wives are more likely to outearn husbands (van Bavel and Klesment 2017) and couples with status dissimilarity (Bratter and Eschbach 2006; Keizer and Komter 2015), marital satisfaction tends to be lower and the likelihood of divorce is higher (Frimmel et al. 2013; Goldstein and Harknett 2006; Schwartz 2010b; Tzeng 1992). All of these findings suggest that the gender reversal in educational attainment and the associated increase in educational hypogamy may, all else equal, contribute to higher levels of marital dissolution.

However, scholars have argued that the risk of divorce among educationally hypogamous couples is not significantly different from other types of educational pairings in countries which have experienced a reversal in the gender gap in higher education (Schwartz and Han 2014; but see Frimmel et al. 2013 for a different pattern in Austria). These findings suggest that the social stigma associated with educational hypogamy has diminished as female college graduates have come to outnumber similarly educated men. Consistent with this hypothesis of changing social norms, one study found that the likelihood of divorce among educationally hypogamous couples is smaller in regions where the proportion of such couples is relatively high (Theunis et al. 2018). But, as van Bavel et al. (2018) correctly note, much research still needs to be done to better understand if, and how, reversal in the gender gap in higher education and the associated increase in educational hypogamy contributes to divorce patterns.

1.1.3.3 Gender Division of Labor

The role of assortative mating in maximizing the gains to marriage is central to the New Home Economics theory of marriage (Becker 1991). To the extent that educational attainment is positively associated with productivity in the labor market (for men) and productivity in domestic labor (for women), we should expect to see relatively high levels of female educational hypergamy (and a pronounced gender division of labor among spouses). Critiques of this "specialization and trading" model of marriage have argued that women's economic contributions to the family are increasingly important and thus increasingly valued in the marriage market (Oppenheimer 1988, 1997). Consistent with this reasoning, several studies have shown that women's earnings are positively associated with both the risk of marriage and the likelihood of homogamy with respect to earnings and educational attainment in the United States (Sweeney 2002; Sweeney and Cancian 2004). In this context, we expect that the more recent rise in educational hypogamy not only reflects societal changes in gender norms, but may also contribute to greater equality within households (Miller 2020). For example, there is much evidence to suggest that women's contribution to household income is positively associated with the time men spend on child care (Raley et al. 2012).

While few studies have directly examined the relationship between educational assortative mating and the gender division of labor, it seems reasonable to posit that educationally hypogamous couples divide their domestic labor, especially child-rearing, more equally than others. There is some evidence that the reverse may be true if women whose husbands' education is lower than their own do more housework to compensate for the deviation from gender status norms (West and Zimmerman 1987), but this pattern of "doing gender" or "deviance neutralization" appears to be relevant to only a small minority of couples (Bittman et al. 2003). Results reported by previous studies are inconsistent. While one analysis of Danish data suggested that educational homogamy is positively associated with husbands' time spent on child care (Bonke and Esping-Andersen 2011), a recent study, using American time-use data, reports that husbands spend more time on child care when they are in an educationally hypogamous marriage, i.e., they have lower educational attainment than their wife (Miller 2020). Yet another study reported that educational assortative mating is unrelated to couples' division of paid labor, measured by the share of work hours, in Sweden and Belgium (Eeckhaut et al. 2014).

1.2 The Aim of the Book

Building on this theoretical foundation, our aims in this manuscript are to situate and evaluate research on Japan in the context of the larger international literature on educational assortative mating. More specifically, we view this as an initial effort to synthesize and understand the societal implications of changes in spouse paring patterns in Japan within the context of existing research on trends and consequences of educational assortative mating in the United States and other low-fertility Western societies. For scholars interested primarily in the demography of Japan, we seek to provide a broad cross-national perspective for thinking about linkages between changes in marriage formation and socioeconomic stratification in Japan. For stratification scholars, we seek to articulate the importance of attention to non-Western societies for a fuller understanding of contextual influences on the social and economic implications of changing patterns of assortative mating. Of particular importance are the facts that Japan continues to be characterized by a sharp gender division of labor in both public and private spheres (Brinton 1993; Estévez-Abe 2008; Nemoto 2016; Osawa 1993; Tsutsui 2020; Yamaguchi 2019; Yoshida 2017) and that it has experienced an increase in income inequality in recent decades (Shirahase 2014; Tachibanaki 2005). Persistent gender inequality means that women's educational attainment is more loosely associated with their labor force participation than is the case in many other wealthy countries (Brinton and Lee 2001; Shambaugh et al. 2017). Japan is also a society that has long been characterized by strong normative expectations of marrying up (hypergamy) (Brinton et al. 2021; Raymo and Iwasawa 2005; Raymo et al. 2015) but has also seen a recent decrease in educational homogamy (Fujihara and Uchikoshi 2019; Fukuda et al. 2021; Miwa 2007).

Our analyses make use of multiple sources of both survey and administrative data to provide a descriptive overview of patterns and trends of "who marries whom" in Japan. We also consider a wide range of potential consequences of trends in educational assortative mating, including income inequality, family formation outcomes, and intergenerational inequality. We discuss our results with reference to the larger international literature.

1.3 An Outline of the Chapters

This book is comprised of this introductory chapter and five additional chapters. In Chap. 2, we provide an overview of prior literature on educational assortative mating. As described earlier, educational assortative mating has a wide range of implications for inequality and social change and we reflect these diverse implications by describing different lines of research on assortative mating. The first section of Chapter 2 focuses on how prior studies have conceptualized assortative mating and its implications. We then discuss explanations for change in patterns of assortative mating.

The literature we review is predominantly based on data from the United States and other Western countries, but we also reference research on a diverse range of institutional contexts, including East Asia and Latin America. Our aim in doing so is to shed light on the importance of institutional characteristics in shaping the nature and the implications of educational assortative mating (Hout and DiPrete 2006). After a discussion of methodological issues in research on assortative mating in Chapter 3, Chapter 4 focuses on the specific context of Japan and how it informs our understanding of patterns, trends, and consequences of educational assortative mating. Chapter 5 is comprised of two sections in which we empirically address questions central to the aims of this book discussed above. First, we document patterns and trends in educational assortative mating in Japan. We also consider the process of marriage, a critical focus in the Japanese context where a rapid decline in marriage, concentrated among particular socioeconomic groups (Fukuda et al. 2020), suggests that the educational composition of married couples differs from that of the general population. Second, we investigate the consequences of educational assortative mating, an important, but understudied, question in Japan where income inequality has increased and research attention to various dimensions of stratification has grown rapidly in recent years. We conclude with Chapter 6, in which we summarize the results of our analyses and discuss future directions, unanswered questions, and novel approaches that could provide new insights.

We conclude this introduction with two caveats about the scope of our focus in this manuscript. First, while the range of theoretically and substantively interesting and important dimensions of assortative mating we might consider is broad (Buss et al. 2001; Buss and Schmitt 2019; Kalmijn 1998), we limit our focus to the widely studied role of educational attainment while occasionally touching on assortative mating by other related characteristics (age and income). Second, we focus on heterosexual couples to avoid complicating the picture with evidence from recent studies examining assortative mating among same-sex couples (Verbakel and Kalmijn 2014). Last, for simplicity, we use the term "mating" rather than "marriage," reflecting the fact that several studies we reference include cohabiting unions.

References

Acemoglu, Daron, and David Autor. 2011. Skills, tasks and technologies: Implications for employment and earnings. *Handbook of Labor Economics* 1043–1171.

Autor, David. 2010. *The polarization of job opportunities in the U.S. labor market*. (Joint report for the Center for American Progress and the Hamilton Project). Retrieved from https://economics.mit.edu/files/5554.

Bai, Yifan. 2018. The consequences of educational assortative marriage for children's academic proficiency in the United States and China. Unpublished Ph.D. Dissertation, Pennsylvania State University.

Beck, Audrey, and Carlos González-Sancho. 2009. Educational assortative mating and children's school readiness. *Center for Research on Child Wellbeing Working Paper* 1–38.

Becker, Gary S. 1991. *A treatise on the family*. Harvard University Press.

Bittman, Michael, Paula England, Liana Sayer, Nancy Folbre, and George Matheson. 2003. When does gender trump money? Bargaining and time in household work. *American Journal of Sociology* 109 (1): 186–214. https://doi.org/10.1086/378341.

Blau, Francine D., and Lawrence M. Kahn. 2007. Changes in the labor supply behavior of married women: 1980–2000. *Journal of Labor Economics* 25 (3): 393–438. https://doi.org/10.1086/513416.

Blossfeld, Hans-Peter. 2009. Educational assortative marriage in comparative perspective. *Annual Review of Sociology* 35 (1): 513–530. https://doi.org/10.1146/annurev-soc-070308-115913.

Boertien, Diederik, and Iñaki Permanyer. 2019. Educational assortative mating as a determinant of changing household income inequality: A 21-country study. *European Sociological Review* 35(4): 522–537.

Bonke, Jens, and Gøsta Esping-Andersen. 2011. Family investments in children-productivities, preferences, and parental child care. *European Sociological Review* 27 (1): 43–55. https://doi.org/10.1093/esr/jcp054.

Brandolini, Andrea, and Timothy M. Smeeding. 2009. Income inequality in richer and OECD countries. In *The Oxford handbook of economic inequality*, ed. B. Nolan, W. Salverda, and T. M. Smeeding, 71–100. Oxford University Press.

Bratter, Jenifer L., and Karl Eschbach. 2006. 'What about the couple?' Interracial marriage and psychological distress. *Social Science Research* 35 (4): 1025–1047. https://doi.org/10.1016/j.ssresearch.2005.09.001.

Breen, Richard, and Signe Andersen. 2012. Educational assortative mating and income inequality in Denmark. *Demography* 49 (3): 867–887. https://doi.org/10.1007/s13524-012-0111-2.

Breen, Richard, and Leire Salazar. 2010. Has increased women's educational attainment led to greater earnings inequality in the United Kingdom? A multivariate decomposition analysis. *European Sociological Review* 26 (2): 143–157. https://doi.org/10.1093/esr/jcp011.

Brinton, Mary C. 1993. *Women and the economic miracle: Gender and work in postwar Japan*. University of California Press.
Brinton, Mary C. 2001. *Women's working lives in East Asia*. Stanford University Press.
Brinton, Mary C., and Sunhwa Lee. 2001. Women's education and the labor market in Japan and South Korea. In *Women's working lives in East Asia*, ed. M. C. Brinton, 125–250. Stanford University Press
Brinton, Mary C., Eunmi Mun, and Ekaterina Hertog. 2021. Singlehood in contemporary Japan: Rating, dating, and waiting for a good match. *Demographic Research* 44: 239–276. https://doi.org/10.4054/DemRes.2021.44.10.
Buss, David M., and David P. Schmitt. 2019. Mate preferences and their behavioral manifestations. *Annual Review of Psychology* 70 (1): 77–110. https://doi.org/10.1146/annurev-psych-010418-103408.
Buss, David M., Todd K. Shackelford, Lee A. Kirkpatrick, and Randy J. Larsen. 2001. A half century of mate preferences: The cultural evolution of values. *Journal of Marriage and the Family* 63 (2): 491–503. https://doi.org/10.1111/j.1741-3737.2001.00491.x.
Byun, Soo-yong, Yifan Bai, and Hee Jin Chung. 2020. Consequences of educational assortative marriage for children's academic achievement in South Korea. In *Korean families yesterday and today*, ed. H. Park and H. Woo, 95–118. University of Michigan Press.
Cancian, Maria, and Deborah Reed. 1998. Assessing the effects of wives' earnings on family income inequality. *The Review of Economics and Statistics* 80 (1): 73–79.
Charles, Kerwin, Erik Hurst, and Alexandra Killewald. 2013. Marital sorting and parental wealth. *Demography* 50 (1): 51–70. https://doi.org/10.1007/s13524-012-0144-6.
Chetty, Raj, Michael Stepner, Sarah Abraham, Shelby Lin, Benjamin Scuderi, Nicholas Turner, Augustin Bergeron, and David Cutler. 2016. The association between income and life expectancy in the United States, 2001–2014. *JAMA* 315 (16): 1750-1766. https://doi.org/10.1001/jama.2016.4226.
Collins, Randall. 1979. *The credential society: An historical sociology of education and stratification*. Academic Press.
Corak, Miles. 2013. Income inequality, equality of opportunity, and intergenerational mobility. *Journal of Economic Perspectives* 27 (3): 79–102. https://doi.org/10.1257/jep.27.3.79.
DiPrete, Thomas A., and Claudia Buchmann. 2013. *The rise of women: The growing gender gap in education and what it means for American Schools*. Russell Sage Foundation.
Dribe, Martin, and Maria Stanfors. 2010. Family life in power couples: Continued childbearing and union stability among the educational elite in Sweden, 1991–2005. *Demographic Research* 23: 847–878. https://doi.org/10.4054/DemRes.2010.23.30.
Duncan, Edited Greg J., and Richard J. Murnane, eds. 2011. *Whither opportunity?: Rising inequality, schools, and children's life chances*. Russell Sage Foundation.
Edwards, Ryan D., and Jennifer Roff. 2016. What mom and dad's match means for junior: Marital sorting and child outcomes. *Labour Economics* 40: 43–56. https://doi.org/10.1016/j.labeco.2016.04.005.
Eeckhaut, Mieke, Maria A. Stanfors, and Bart Van de Putte. 2014. Educational heterogamy and the division of paid labour in the family: A comparison of present-day Belgium and Sweden. *European Sociological Review* 30 (1): 64–75. https://doi.org/10.1093/esr/jct022.
Eika, Lasse, Magne Mogstad, and Basit Zafar. 2019. Educational assortative mating and household income inequality. *Journal of Political Economy* 127 (6): 1–41.
Esping-Andersen, Gøsta. 2007. Sociological explanations of changing income distributions. *American Behavioral Scientist* 50 (5): 639–658.
Esping-Andersen, Gøsta., and Francesco C. Billari. 2015. Re-theorizing family demographics. *Population and Development Review* 41 (1): 1–31. https://doi.org/10.1111/j.1728-4457.2015.00024.x.
Esteve, Albert, Joan Garcia-Roman, and Inaki Permanyer. 2012. The gender-gap reversal in education and its effect on union formation: The end of hypergamy? *Population and Development Review* 38 (3): 535–546. https://doi.org/10.1111/j.1728-4457.2012.00515.x.

Esteve, Albert, Christine R. Schwartz, Jan van Bavel, Inaki Permanyer, Martin Klesment, and Joan Garcia-Roman. 2016. The end of hypergamy: Global trends and implications. *Population and Development Review* 42 (4): 615–625.

Estévez-Abe, Margarita. 2008. *Welfare and capitalism in postwar Japan.* Cambridge University Press.

Frimmel, Wolfgang, Martin Halla, and Rudolf Winter Ebmer. 2013. Assortative mating and divorce: Evidence from Austrian Register Data. *Journal of the Royal Statistical Society: Series A (statistics in Society)* 176 (4): 907–929.

Fujihara, Sho, and Fumiya Uchikoshi. 2019. Declining association with persistent gender asymmetric structure: Patterns and trends in educational assortative marriage in Japan, 1950–1979. *Research in Social Stratification and Mobility* 60: 66–77. https://doi.org/10.1016/j.rssm.2018.12.001.

Fukuda, Setsuya, Shohei Yoda, and Ryohei Mogi. 2021. Educational assortative mating in Japan: Evidence from the 1980–2010 census. *The Journal of Population Studies (Jinkogaku Kenkyu).* https://doi.org/10.24454/jps.2101001.

Fukuda, Setsuya, James M. Raymo, and Shohei Yoda. 2020. Revisiting the educational gradient in marriage in Japan. *Journal of Marriage and Family* 82 (4): 1378–1396. https://doi.org/10.1111/jomf.12648.

Goldin, Claudia Dale, and Lawrence F. Katz. 2009. *The race between education and technology.* Harvard University Press.

Goldscheider, Frances, Eva Bernhardt, and Trude Lappegård. 2015. The gender revolution: A framework for understanding changing family and demographic behavior. *Population and Development Review* 41 (2): 207–239. https://doi.org/10.1111/j.1728-4457.2015.00045.x.

Goldstein, Joshua R., and Kristen Harknett. 2006. Parenting across racial and class lines: Assortative mating patterns of new parents who are married, cohabiting, dating or no longer romantically involved. *Social Forces* 85 (1): 121–143.

Gonalons-Pons, Pilar, and Christine R. Schwartz. 2017. Trends in economic homogamy: Changes in assortative mating or the division of labor in marriage? *Demography* 54 (3): 985–1005. https://doi.org/10.1007/s13524-017-0576-0.

Gornick, Janet C., and Markus Jäntti. 2012. Child poverty in cross-national perspective: Lessons from the Luxembourg Income Study. *Children and Youth Services Review* 34 (3): 558–568. https://doi.org/10.1016/j.childyouth.2011.10.016.

Greenwood, Jeremy, Nezih Guner, Georgi Kocharkov, and Cezar Santos. 2014. Marry your like: Assortative mating and income inequality. *The American Economic Review* 104 (5): 1–15.

Greenwood, Jeremy, Nezih Guner, Georgi Kocharkov, and Cezar Santos. 2015. Corrigendum to marry your like: Assortative mating and income inequality. http://www.Jeremygreenwood.Net/Papers/GgksPandPcorrigendum.Pdf.

Herzberg-Druker, Efrat, and Haya Stier. 2019. Family matters: The contribution of households' educational and employment composition to income inequality. *Social Science Research* 82: 221–239. https://doi.org/10.1016/j.ssresearch.2019.04.012.

Hout, Michael, and Thomas A. DiPrete. 2006. What we have learned: RC28's Contributions to knowledge about social stratification. *Research in Social Stratification and Mobility* 24 (1): 1–20. https://doi.org/10.1016/j.rssm.2005.10.001.

Hryshko, Dmytro, Chinhui Juhn, and Kristin McCue. 2017. Trends in earnings inequality and earnings instability among U.S. couples: How important is assortative matching? *Labour Economics* 48: 168–182. https://doi.org/10.1016/j.labeco.2017.08.006.

Hu, Anning, and Zhenchao Qian. 2015. Educational homogamy and earnings inequality of married couples: Urban China, 1988–2007. *Research in Social Stratification and Mobility* 40: 1–15. https://doi.org/10.1016/j.rssm.2015.01.004.

Kalmijn, Matthijs. 1991. Shifting boundaries: Trends in religious and educational homogamy. *American Sociological Review* 56 (6): 786–800.

Kalmijn, Matthijs. 1998. Intermarriage and homogamy: Causes, patterns, trends. *Annual Review of Sociology* 24: 395–421.

Keizer, R., and A. Komter. 2015. Are 'Equals' happier than 'Less Equals'? A couple analysis of similarity and well-being. *Journal of Marriage and Family* 77(4): 954–967.
Killewald, Alexandra, and Xiaolin Zhuo. 2019. U.S. Mothers' long-term employment patterns. *Demography* 56 (1): 285–320. https://doi.org/10.1007/s13524-018-0745-9.
Kollmeyer, Christopher. 2013. Family structure, female employment, and national income inequality: A cross-national study of 16 Western countries. *European Sociological Review* 29 (4): 816–827. https://doi.org/10.1093/esr/jcs060.
Lundberg, Shelly, Robert A. Pollak, and Jenna Stearns. 2016. Family inequality: Diverging patterns in marriage, cohabitation, and childbearing. *Journal of Economic Perspectives* 30 (2): 79–102. https://doi.org/10.1257/jep.30.2.79.
Mare, Robert D. 2016. Educational homogamy in two gilded ages: Evidence from inter-generational social mobility data. *The ANNALS of the American Academy of Political and Social Science* 663 (1): 117–139. https://doi.org/10.1177/0002716215596967.
McCall, Leslie, and Christine Percheski. 2010. Income inequality: New trends and research directions. *Annual Review of Sociology* 36 (1): 329–347. https://doi.org/10.1146/annurev.soc.012809.102541.
McLanahan, Sara, and Christine Percheski. 2008. Family structure and the reproduction of inequalities. *Annual Review of Sociology* 34 (1): 257–276. https://doi.org/10.1146/annurev.soc.34.040507.134549.
Miller, Rhiannon N. 2020. Educational assortative mating and time use in the home. *Social Science Research* 90:102440. https://doi.org/10.1016/j.ssresearch.2020.102440.
Miwa, Satoshi. 2007. Long-term trends in status homogamy. In *Deciphering stratification and inequality*, ed. Y. Sato, 140–60. Trans Pacific Press.
Monaghan, David. 2015. Income inequality and educational assortative mating: Evidence from the Luxembourg income study. *Social Science Research* 52: 253–269. https://doi.org/10.1016/j.ssresearch.2015.02.001.
Murphy, Raymond. 1984. The structure of closure: A critique and development of the Theories of Weber, Collins, and Parkin. *The British Journal of Sociology* 35 (4): 547–567. https://doi.org/10.2307/590434.
Nemoto, Kumiko. 2016. *Too few women at the top: The persistence of inequality in Japan*. ILR Press, an imprint of Cornell University Press.
Nitsche, Natalie, Anna Matysiak, Jan van Bavel, and Daniele Vignoli. 2018. Partners' educational pairings and fertility across Europe. *Demography* 55 (4): 1195–1232. https://doi.org/10.1007/s13524-018-0681-8.
Nomes, Eli, and Jan van Bavel. 2016. Marital fertility and educational assortative mating before, during, and after the baby boom in Belgium. *Annales De Démographie Historique* 132 (2): 139–163.
Oppenheimer, Valerie Kincade. 1988. A theory of marriage timing. *American Journal of Sociology* 94 (3): 563–591.
Oppenheimer, Valerie Kincade. 1997. Women's employment and the gain to marriage: The specialization and trading model. *Annual Review of Sociology* 23: 431–453.
Osawa, Machiko. 1993. *Keizai Henka to Joshi Rodo (Economic Change and Women's Work)*. Nihon Keizai Shimbunsha.
Osiewalska, Beata. 2017. Childlessness and fertility by couples' educational gender (in)Equality in Austria, Bulgaria, and France. *Demographic Research* 37: 325–362. https://doi.org/10.4054/DemRes.2017.37.12.
Parkin, Frank. 1971. *Class inequality and political order*. University of California Press.
Pesando, Luca Maria. 2021. A four-country study on the relationship between parental educational homogamy and children's health from infancy to adolescence. *Population Research and Policy Review*. https://doi.org/10.1007/s11113-020-09627-2.
Piketty, Thomas. 2014. *Capital in the twenty-first century*. Harvard University Press.

Qian, Yue. 2018. Educational assortative mating and income dynamics in couples: A longitudinal dyadic perspective. *Journal of Marriage and Family* 80 (3): 607–621. https://doi.org/10.1111/jomf.12470.

Raley, Sara, Suzanne M. Bianchi, and Wendy Wang. 2012. When do fathers care? Mothers' economic contribution and fathers' involvement in child care. *American Journal of Sociology* 117 (5): 1422–1459. https://doi.org/10.1086/663354.

Rauscher, Emily. 2020. Why who marries whom matters: Effects of educational assortative mating on infant health in the United States 1969–1994. *Social Forces* 98 (3): 1143–1173. https://doi.org/10.1093/sf/soz051.

Raymo, James M., and Miho Iwasawa. 2005. Marriage market mismatches in Japan: An alternative view of the relationship between women's education and marriage. *American Sociological Review* 70 (5): 801–822. https://doi.org/10.1177/000312240507000504.

Raymo, James M., Hyunjoon Park, Yu Xie, and Wei-jun Jean Yeung. 2015. Marriage and family in East Asia: Continuity and change. *Annual Review of Sociology* 41: 471–92. https://doi.org/10.1146/annurev-soc-073014-112428.

Rosenfeld, Michael J. 2008. Racial, educational and religious endogamy in the United States: A comparative historical perspective. *Social Forces* 87 (1): 1–31.

Schwartz, Christine R. 2010a. Earnings inequality and the changing association between spouses' earnings. *American Journal of Sociology* 115 (5): 1524–1557. https://doi.org/10.1086/651373.

Schwartz, Christine R. 2010b. Pathways to educational homogamy in marital and cohabiting unions. *Demography* 47 (3): 735–753.

Schwartz, Christine R. 2013. Trends and variation in assortative mating: Causes and consequences. *Annual Review of Sociology* 39 (1): 451–470. https://doi.org/10.1146/annurev-soc-071312-145544.

Schwartz, Christine R., and Hongyun Han. 2014. The reversal of the gender gap in education and trends in marital dissolution. *American Sociological Review* 79 (4): 605–629. https://doi.org/10.1177/0003122414539682.

Seltzer, Nathan. 2019. Beyond the great recession: labor market polarization and ongoing fertility decline in the United States. *Demography* 56 (4): 1463–1493. https://doi.org/10.1007/s13524-019-00790-6.

Shafer, Emily Fitzgibbons. 2011. Wives' relative wages, husbands' paid work hours, and wives' labor-force exit. *Journal of Marriage and Family* 73 (1): 250–263. https://doi.org/10.1111/j.1741-3737.2010.00802.x.

Shambaugh, Jay, Ryan Nunn, and Becca Portman. 2017. Lessons from the rise of women's labor force participation in Japan. *The Brookings Institution.*

Shirahase, Sawako. 2014. *Social inequality in Japan.* Routledge.

Smits, Jeroen, Wout Ultee, and Jan Lammers. 1998. Educational homogamy in 65 countries: An explanation of differences in openness using country-level explanatory variables. *American Sociological Review* 63 (2): 264–285. https://doi.org/10.2307/2657327.

Sweeney, Megan M. 2002. Two decades of family change: The shifting economic foundations of marriage. *American Sociological Review* 67 (1): 132–147. https://doi.org/10.2307/3088937.

Sweeney, Megan M., and Maria Cancian. 2004. The changing importance of White Women's economic prospects for assortative mating. *Journal of Marriage and Family* 66 (4): 1015–1028. https://doi.org/10.1111/j.0022-2445.2004.00073.x.

Tachibanaki, Toshiaki. 2005. *Confronting income inequality in Japan.* MIT Press.

Theunis, Lindsay, Christine Schnor, Didier Willaert, and Jan van Bavel. 2018. His and her education and marital dissolution: Adding a contextual dimension. *European Journal of Population* 34 (4): 663–687. https://doi.org/10.1007/s10680-017-9448-y.

Torche, Florencia. 2010. Educational assortative mating and economic inequality: A comparative analysis of three Latin American countries. *Demography* 47 (2): 481–502.

Treas, Judith. 1987. The effect of women's labor force participation on the distribution of income in the United States. *Annual Review of Sociology* 13: 259–288.

Treiman, Donald J. 1970. Industrialization and social stratification. *Sociological Inquiry* 40 (2): 207–234.
Trimarchi, Alessandra, and Jan van Bavel. 2020. Partners' educational characteristics and fertility: Disentangling the effects of earning potential and unemployment risk on second births. *European Journal of Population* 36 (3): 439–464. https://doi.org/10.1007/s10680-019-09537-w.
Tsutsui, Junya. 2020. *Work and family in Japanese society*. Springer Singapore.
Tzeng, Meei-Shenn. 1992. The effects of socioeconomic heterogamy and changes on marital dissolution for first marriages. *Journal of Marriage and Family* 54 (3): 609–619.
Uchikoshi, Fumiya. 2016. Intergenerational succession of educational assortative mating and its trends: An analysis of integrated large-scale survey data. *Japanese Journal of Family Sociology* 28(2):136–147 (in Japanese). https://doi.org/10.4234/jjoffamilysociology.28.136.
Uchikoshi, Fumiya. 2019. Consequences of educational assortative mating through divorce. *Japanese Sociological Review* 70 (1): 10–26 (in Japanese). https://doi.org/10.4057/jsr.70.10.
van Bavel, Jan. 2012. The reversal of the gender gap in education and female breadwinners in Europe ed. by W. Lutz, K. S. James, V. Skirbekk, and J. van Bavel. *Vienna Yearbook of Population Research* 10: 127–154. https://doi.org/10.1553/populationyearbook2012s127.
van Bavel, Jan and Martin Klesment. 2017. Educational pairings, motherhood, and women's relative earnings in Europe. *Demography* 54 (6): 2331–2349. https://doi.org/10.1007/s13524-017-0621-z.
van Bavel, Jan, Christine R. Schwartz, and Albert Esteve. 2018. The reversal of the gender gap in education and its consequences for family life. *Annual Review of Sociology* 44 (1): 341–360. https://doi.org/10.1146/annurev-soc-073117-041215.
Wagner, Sander, Diederik Boertien, and Mette Gørtz. 2020. The wealth of parents: Trends over time in assortative mating based on parental wealth. *Demography* 57 (5): 1809–1831
West, Candace, and Don H. Zimmermann. 1987. Doing gender. *Gender and Society* 12: 125–151.
Western, Bruce, Deirdre Bloome, and Christine Percheski. 2008. Inequality among American families with children, 1975 to 2005. *American Sociological Review* 73 (6): 903–920. https://doi.org/10.1177/000312240807300602.
Yamaguchi, Kazuo. 2019. *Gender inequalities in the Japanese workplace and employment: Theories and empirical evidence*. Vol. 22. Springer Singapore.
Yoshida, Akiko. 2017. *Unmarried women in Japan*. Routledge.
Zagel, Hannah, and Richard Brreen. 2019. Family demography and income inequality in West Germany and the United States. *Acta Sociologica* 62 (2): 174–192. https://doi.org/10.1177/0001699318759404.

Chapter 2
Conceptual Issues

2.1 Introduction

In this chapter, we discuss two important conceptual issues regarding assortative mating, one about why marriages tend to be homogamous and the other about why patterns of assortative mating may change over time. Drawing on prior literature, we review three key factors thought to shape the assortativity of marriage—preference, opportunity, and third-party influence. We also touch on two critical sorting mechanisms that shape patterns of assortative mating—status exchange and winnowing. Our discussion of explanations for changes in assortative mating begins with a focus on modernization theory. This important perspective emphasizes the roles of structural changes in industrialized society as well as educational expansion. In addition to this macro-level explanation of the change, we also review theoretical emphases on how the changing economic landscape impacts the marriage market, paying particular attention to women's growing earnings potential. Finally, we touch upon supply-side mechanisms and how changes in the pool of marriageable men and women in the marriage market influence patterns of assortative mating.

2.2 Explanations for Assortative Mating

2.2.1 *Preference*

If we consider marriage as both an individual and a joint decision, it is clear that preferences influence whom people expect to partner with. As mentioned above, the specifics of mate preferences can vary widely, from physical characteristics (e.g., height) or demographic characteristics (e.g., age) to socioeconomic status (e.g., income) (Buss and Schmitt 2019). Among these various traits, previous studies in the social scientific literature have, for several reasons, paid particular attention to the role of education. First, marriage typically occurs relatively early in the life course, when

F. Uchikoshi and J. M. Raymo, *Educational Assortative Mating in Japan*,
Population Studies of Japan, https://doi.org/10.1007/978-981-16-3713-1_2

the long-term value of socioeconomic characteristics that tend to fluctuate over time (e.g., occupational status and earnings) is difficult to accurately assess from current characteristics or everyday interactions (Oppenheimer 1988). Compared with these measures, educational attainment is a good predictor of future earnings that is relatively easy to observe or anticipate in young adulthood. Second, educational attainment serves not only as a signal of future economic potential but is also correlated with cultural tastes and values relevant to the spouse selection process (Blossfeld 2009; Kalmijn 1998). Third, given the temporal proximity of marriage to school completion, educational settings play an important role in providing opportunities to meet partners (Mare 1991). Fourth, practically speaking, educational attainment is relatively easy to observe and accurately measure for both men and women (and researchers). Income and wealth not only fluctuate over time but are also measured with error (including high levels of non-response) on social surveys. Occupation also fluctuates over time and earlier studies of occupational homogamy struggled with the fact that, historically, many women did not have regular employment prior to marriage (Hout 1982). Compared with these other theoretically relevant indicators of socioeconomic status, relatively high response rates and less measurement error have resulted in educational attainment being the most widely studied dimension of socioeconomic assortative mating (Blossfeld 2009).

It is possible that the key signals derived from educational attainment in the marriage market may vary within the population. For example, some may view educational characteristics primarily as a predictor of future earnings (Oppenheimer 1988), while others may view educational attainment primarily as a signal of tastes, values, and attitudes (Kalmijn 1994, 1998). In either case, preferences of highly educated singles for similarly educated spouses will result in educational homogamy assuming that there are no structural barriers in the marriage market to the formation of such marriages (e.g., spatial or institutional segregation).

Previous research has often assumed, either explicitly or implicitly, that socioeconomic characteristics are valued differently in the marriage market by men and women. One classic example is the theoretical emphasis in economic models of marriage (e.g., Becker 1991) on the gains to marriage derived via role specialization in which one partner focuses on market labor while the other focuses on domestic labor. Although this theory is gender-neutral, empirically it is clear that men have tended to specialize in the breadwinner role and women in the homemaker role. To the extent that educational attainment is more strongly associated with productivity in employment than in domestic work, this theory of marriage is consistent with a relatively high prevalence of educationally hypergamous marriages. Understanding this pattern of spouse pairing as a reflection of individual choices directed at maximizing the gains to marriage has led scholars to speculate that the prevalence of educational hypergamy (women marrying up) reflects women's preference to marry up and men's preference to marry less-educated women. Recent studies, however, find convergence in men's and women's mate preferences (Buss et al. 2001), which might underlie the recently observed increase in educational hypogamy (women marrying down) (Esteve et al. 2012, 2016; van Bavel 2012).

2.2.2 Opportunity Structure

Spouse pairing patterns are also constrained by local marriage market structure (Lewis and Oppenheimer 2000; Lichter et al. 1995). The vague concept of a marriage "market" reflects the fact that individuals can meet potential partners in a variety of social settings, including school, workplace, neighborhood, and friendship networks. Because the distribution of specific individual characteristics varies by setting and can change over time, *who meets where* influences *who marries whom* (McClendon 2018). One straightforward example is that of religious assortative mating, with those who are active in communities of individuals that share the same faith (Kalmijn and Flap 2001) or who live in regions with a higher concentration of coreligionists (McClendon 2016) more likely to marry someone of the same religion. Similarly, couples who meet while in school tend to be more educationally homogamous, and the likelihood of educational homogamy tends to decrease with time since the completion of education (Mare 1991; Shafer and Qian 2010).

2.2.3 Third-Party Influence

Theories about preferences or opportunities are more or less "individualistic" in the sense that they assume that marriage markets in contemporary societies are defined primarily with respect to individuals' achieved status (Kalmijn 1998). A small number of studies, however, has argued that "third actors" can play an important role in explaining spouse pairing patterns (Charles et al. 2013; Henz and Mills 2018; Hu 2016; Kalmijn 1991a; Kalmijn and Flap 2001; Mäenpää 2015; Tian and Davis 2019; Uunk et al. 1996). As Blau and Duncan (1967: 358) compellingly argued: "parents can encourage their children to marry spouses with the proper family background ... by placing their children into an environment largely populated by prospective mates with suitable family connections." Thus, parents, or social origins more generally, may still be of importance, even in the marriage markets of wealthier societies (Armstrong and Hamilton 2021; Kalmijn 1998; Musick et al. 2012). Empirical evidence for this can be found in differing levels of homogamy with respect to traits such as religion within the same population (Kalmijn 1998) or class privilege among college graduates (Armstrong and Hamilton 2021; Musick et al. 2012). Having said that, it is clear that parental control over children's spouse selection has decreased over time and that patterns of assortative mating largely reflect decisions independent of parental/family influence (Rosenfeld and Kim 2005; Rosenfeld and Thomas 2012).

Instead of parents influencing their children's marriage, children may also take advantage of their parents' status to shape their own position in the marriage market (Musick et al. 2012). Prior studies suggest that this may have been especially true in the past for women who faced structural barriers to accessing higher education and career employment. In settings where women's future economic position depends less on their own achieved SES than that of their husbands (Sorensen 1994), women from higher SES backgrounds had an incentive to rely on their parents' socioeconomic status to marry a spouse with higher SES (Schwartz et al. 2016). This gender difference in the strategic use of parents' characteristics to maximize economic well-being has been documented in several studies (Blackwell 1998; Chadwick and Solon 2002), although other research reports no gender difference (Schwartz et al. 2016) and we expect this pattern to have waned over time in response to convergence in men's and women's educational attainment and occupational characteristics.

2.2.4 Status Exchange

Status exchange theory focuses on the process in the marriage market through which individuals exchange their relatively advantaged social status (e.g., university degree) to offset their relatively disadvantaged status (e.g., racial minority status) (Davis 1941; Merton 1941). The process of status exchange should thus increase heterogamous pairings. The most well-known and widely studied example of status exchange in the United States is intermarriage of high-SES racial minorities and low-SES whites (Fu 2010; Gullickson and Fu 2010; Kalmijn 1993, 2010; Qian 1997; Rosenfeld 2005, 2010). An interesting recent study of the online dating market suggests that patterns of status exchange may be different in that setting, with white men and women more likely to interact with each other regardless of educational attainment and less likely to date highly educated racial minorities (Lin and Lundquist 2013). Other examples of status exchange include, but are not limited to, achieved SES and participation in high culture like music or art events (DiMaggio and Mohr 1985), marital status and race (Choi and Tienda 2017; Qian 2018, but see Fu 2010 for different findings), and caste and education (Lin et al. 2020).

2.2.5 Winnowing

Although most empirical analyses of assortative mating focus on spouse pairing patterns at a given point in time, it is important to emphasize that assortative mating can be conceptualized as a dynamic process in which individuals are sorted into marriage (Kalmijn 1998; Schwartz and Mare 2012). The so-called winnowing hypothesis highlights the importance of this dynamic sorting process. Winnowing refers to a process in which the selective dissolution of cohabiting unions works to "sweep out" partners with undesirable characteristics (Blackwell and Lichter 2000, 2004) and results in couples becoming more homogamous in terms of socioeconomic traits. However, evidence on the winnowing hypothesis is inconsistent. Prior studies suggest that this process is especially relevant for interracial relationships (Blackwell and Lichter 2004; Wang et al. 2006), while other studies have reported that resemblance by racial (Qian and Lichter 2007) and educational (Goldstein and

Harknett 2006; Schwartz 2010b) traits does not differ between cohabiting couples who split up and those who got married (Schwartz 2013).

Despite these empirical inconsistencies, the winnowing hypothesis, or depictions of assortative mating as a dynamic process in general, provides important insights into the mechanisms of marital sorting (Grow and van Bavel 2015; Lin and Lundquist 2013; Xie et al. 2015). Detailed analysis of the process of mate selection, for example, allows us to understand when and how educational attainment or other relevant traits play a role in matching processes. Focusing on dynamic aspects of mate selection also helps us to think about the important empirical question of how to observe and measure the process of assortative mating. Because surveys, either cross-sectional or longitudinal, typically cannot track unrelated individuals who happen to meet and marry, observation of the winnowing process is not easily accomplished. However, there are a few informative studies, such as those focusing on the online dating market to examine the reciprocal process of mate selection (Bruch and Newman 2018, 2019; Lin and Lundquist 2013) in a way that allows us to track unrelated individuals' dating experiences and to quantify the selection process at the population level (Schwartz and Mare 2012). These online dating studies have revealed the regularity of preferences for pairing with a more "desirable" partner. For example, one study found that both men and women tend to send messages to potential mates who are about 25% more desirable than themselves, based on a desirability score reflecting the number of messages they received (Bruch and Newman 2018). At the same time, another study found that the likelihood of receiving a message reflects the preexisting racial hierarchy (Lin and Lundquist 2013), with racial minorities less likely to receive messages from potential dates. Given that matching involves interactions between two actors, the asymmetric relationship between sending and receiving interest suggests that knowing one's partner preference (or quantifying the marriage market availability) is not enough to capture the reality of the matching process. Rather, future studies need to develop tools for investigating the reciprocal relationships of mating preferences and how they lead to the observed matching patterns.

2.3 Why Assortative Mating Changes

2.3.1 Modernization and Educational Expansion

In their summaries of earlier research, Blossfeld and Timm (2003) and Rauscher (2015) emphasize that temporal changes in patterns of educational assortative mating depend on how institutional contexts influence preferences and opportunity structure in the marriage market (Kalmijn 1998). The expansion of higher education may be particularly relevant for understanding how both of these mechanisms shape trends in who marries whom.

First, modernization theory posits that educational attainment will replace ascribed characteristics as a key spouse selection criterion in meritocratic societies. Empirical support for this prediction of modernization theory can be found in studies showing a decline in assortative mating in the United States with respect to measures of ascribed status, such as parental occupation (Bouchet-Valat 2014; Kalmijn 1991a), race (Qian and Lichter 2007; Rosenfeld 2008; but see Lichter et al. 2011 and Qian and Lichter 2007 for evidence of increasing homogamy among Hispanics), or religion (Kalmijn 1991b; Rosenfeld 2008). Similarly, Kalmijn (1991a) demonstrated that homogamy with respect to achieved status, measured by educational attainment, has increased over time (see also Mare 1991; Schwartz and Mare 2005).

Second, structural theory (Blau and Schwartz 1984; Rytina et al. 1988) also posits that educational expansion increases educational homogamy. According to this theory, "as group size increases, the probable rate of outgroup relations decreases" (Blau and Schwartz 1984: 31): that is, relative group size per se is seen as a determinant of the possibility of meeting spouses in the same or different social groups. Specifically, this theory predicts that expansion of higher education increases educational homogamy, especially among the highly educated, because such expansion increases the possibility of meeting equally educated spouses at the ages when marriage is most likely to occur (Blossfeld 2009; Mare 1991). Hu and Qian's (2016) recent study of urban China provides evidence consistent with this hypothesis.

Other hypotheses focused on preferences, rather than opportunity structure in the marriage market, also predict that educational expansion leads to increased homogamy. The status attainment hypothesis, for example, predicts that socioeconomic development brings about a change from ascriptive to universalistic achievement criteria, whereby social origins are replaced by educational attainment as the main determinant of one's future socioeconomic status (Blossfeld and Timm 2003; Kalmijn 1998; Treiman 1970). Individuals who achieve high socioeconomic status tend to take education level into account when selecting their partners (Smits et al. 1998). Thus, this hypothesis predicts that the level of educational homogamy increases as individuals increasingly value education in the marriage market as a result of educational expansion.[1]

In contrast, Smits (2003) and Smits and Park (2009) argue that educational expansion should result in declining educational homogamy. According to what they call the *exclusivity* hypothesis, educational homogamy among college graduates is stronger when their group size is smaller. The rationale for this prediction comes from theories of status closure and status-group credentialism (Brown 2001), which posit that small, elite groups' awareness of their advantages motivates efforts to maintain social boundaries by excluding outgroups (Parkin 1971). If group size increases, however, the relative value of higher education decreases, and barriers to mating with members of elite groups should also decrease. Therefore, this hypothesis predicts

[1] Although we focus on the role of educational expansion, we are aware that spouse pairing is not an "either-or" process. As earlier studies have noted (Kalmijn 1998; Schwartz et al. 2016), pairing is a multivariate, not a univariate, process, with individuals having preferences for multiple traits that change over time. Observed patterns of assortative mating with respect to any one characteristic (or combination of characteristics) are thus the result of the complex interplay of these processes.

that educational expansion should promote intermarriage between educational elites and others.

2.3.2 Shifting Economic Foundations of Marriage and Gender Contexts

A similar, but qualitatively different, argument emphasizes the role of increasing economic sorting (Schwartz and Mare 2005). According to the "shifting economic foundations of marriage" thesis (Oppenheimer 1988; Oppenheimer et al. 1997; Sweeney 2002), increasing economic precarity contributes to later and less marriage among those with lower levels of education, while the increasing valuation of women's financial contributions to the family results in later, but higher, levels of marriage among the more educated (Goldstein and Kenny 2001; Kalmijn 2013). This also implies increasing homogamy among high-earning, highly educated partners as the increase of women's contribution to household income reduces men's preference for educational hypergamy (Schwartz and Mare 2005: 642; South 1991; van Bavel 2012). This hypothesis has been supported with data from countries where women's access to both education and the labor market improved in the twentieth century (Han 2010; Ravazzini et al. 2017; Wong 2003).

More broadly, this argument echoes the gender revolution theory, which posits that structural changes in gender relationships within the public and private spheres play a key role in shifting socioeconomic gradients in family outcomes (Goldscheider et al. 2015). Specifically, Goldscheider et al. (2015) argued that there has been an increase in men's involvement in housework and child care, which has contributed to stabilizing families and increasing fertility. Although this theory primarily focuses on fertility as an outcome, it also suggests that, as both men and women come to possess more egalitarian attitudes about the gender division of labor, certain types of spouse pairing patterns may become more common. In particular, it suggests an increase in educational hypogamy and a decline in educational hypergamy—a pairing that has been traditionally associated with gender specialization (Becker 1991). It is important to note, however, that an increase in educational hypogamy does not necessarily imply an increase in couples comprised of two high-income earners (Chudnovskaya and Kashyap 2020) or in women's labor force participation (Lin et al. 2020). Indeed, it is often the case that educational hypogamy is a result of women marrying up in income (Qian and Qian 2017)—a contemporary example of the exchange process described above.

Other scholars, however, have suggested that preferences for mating patterns consistent with the traditional male breadwinner and female homemaker model remain strong (van Bavel et al. 2018). One explanation for the slow convergence of mating preferences despite the rapid increase in women's access to higher education can be found in discussions of the stalled gender revolution (England 2010). According to this literature, the mechanism underlying the observed stalling of gender convergence is persistent gender essentialism: that is, tacit ideas related to fundamental differences between men and women that are deeply rooted in the social structure (England 2010; Ridgeway 2001; Ridgeway and Correll 2004). Scholars suggest that persistent gender essentialism results in gender-asymmetric change (Ridgeway 2011) such that men's shift to gender equality is slower than that of women in terms of the division of household labor (Bianchi et al. 2012), college major choice (England and Li 2006), and preference for educational hypergamy (van Bavel et al. 2018).

2.3.3 Changing Opportunities

Even if individuals' pairing preferences are constant across time, compositional changes in the pool of potential spouses can result in changing spouse pairing patterns. A well-known example is marriage squeezes generated by baby booms (Akers 1967; Muhsam 1974; Schoen 1983) or a "shortage of marriageable men" generated by declining economic opportunities and rising incarceration rates (Lichter et al. 1992; Lichter et al. 2020; Raley 1996; Western 2006). Related to educational assortative mating, studies have argued that relative improvements in women's education (decline in the relative supply of highly educated men) lead to marriage market mismatches detrimental to both highly educated women and less-educated men in societies characterized by strong preferences for female status hypergamy, such as Japan (Raymo and Iwasawa 2005) and South Korea (Raymo and Park 2020). Such marriage market mismatches are particularly relevant for previously married individuals (Qian and Lichter 2018) and may partially account for the distinctive pattern of educational assortative mating among remarried couples (Shafer 2013).

Changes in the pool of potential spouses also reflect shifting opportunities for unmarried men and women to meet in the marriage market. For example, the trend toward later marriage increases the time gap between school completion and marriage, thus contributing to changes in educational assortative mating (Halpin and Chan 2003; Hou and Myles 2008). Mare (1991) hypothesized that the odds of crossing educational barriers (i.e., educational heterogamy) are positively associated with the time gap. At any given average age of marriage, an increase in average educational attainment narrows the time gap, shortens the period for seeking a potential partner, and thus contributes to a higher likelihood that individuals meet their partners in school. In contrast, an increase in the average age at marriage extends the time gap and provides individuals with increased opportunities to meet their partners in settings other than school, such as the workplace, thus leading to more educational heterogamy (Kalmijn 1998).

It is also likely that other recent, fundamental shifts in opportunities to meet are contributing to shifting patterns of educational assortative mating. The most dramatic change is the rise in opportunities to meet online, particularly through dating apps (Rosenfeld 2007). Although about one-third of US adults now meet their intimate partner online (Rosenfeld et al. 2019), with clear implications for certain types of mating (e.g., same-sex pairing, Rosenfeld 2007), its impact on educational assortative mating is still unclear. For example, Thomas (2019) and Potarca (2017) conclude that meeting online is associated with more educational heterogamy in the United States, but Lee's (2016) analysis of online dating data shows that meeting online increases educational homogamy. It is thus possible that sorting mechanisms may differ across online settings and the purposes for which they are used (Bruch and Newman 2018; 2019), with dating apps used to select partners with similar education, while engagement in other types of online communities may result in more educationally heterogamous encounters.

References

Akers, Donald S. 1967. On measuring the marriage squeeze. *Demography* 4 (2): 907–924. https://doi.org/10.2307/2060328.

Armstrong, Elizabeth A., and Laura T. Hamilton. 2021. Classed pathways to marriage: Hometown ties, college networks, and life after graduation. *Journal of Marriage and Family*. https://doi.org/10.1111/jomf.12747.

Becker, Gary S. 1991. *A treatise on the family*. Harvard University Press.

Bianchi, Suzanne M., Liana C. Sayer, Melissa A. Milkie, and John P. Robinson. 2012. Housework: Who did, does or will do it, and how much does it matter? *Social Forces* 91 (1): 55–63. https://doi.org/10.1093/sf/sos120.

Blackwell, Debra L. 1998. Marital homogamy in the United States: The influence of individual and paternal education. *Social Science Research* 27 (2): 159–188. https://doi.org/10.1006/ssre.1998.0618.

Blackwell, Debra L., and Daniel T. Lichter. 2000. Mate selection among married and cohabiting couples. *Journal of Family Issues* 21 (3): 275–302. https://doi.org/10.1177/019251300021003001.

Blackwell, Debra L., and Daniel T. Lichter. 2004. Homogamy among dating, cohabiting, and married couples. *Sociological Quarterly* 45 (4): 719–737. https://doi.org/10.1525/tsq.2004.45.4.719.

Blau, Peter Michael, and Otis Dudley Duncan. 1967. *The American occupational structure*. Wiley.

Blau, Peter M., and Joseph E. Schwartz. 1984. *Crosscutting social circles: Testing a macrostructural theory of intergroup relations*. Academic Press.

Blossfeld, Hans-Peter. 2009. Educational assortative marriage in comparative perspective. *Annual Review of Sociology* 35 (1): 513–530. https://doi.org/10.1146/annurev-soc-070308-115913.

Blossfeld, Hans-Peter., and Andreas Timm. 2003. *Who marries whom?* Springer.

Bouchet-Valat, Milan. 2014. Changes in educational, social class and social class of origin homogamy in France (1969–2011): Greater openness overall but increased closure of elites. *Revue Française De Sociologie* 55 (3): 324–364.

Brown, David K. 2001. The social sources of educational credentialism: Status cultures, labor markets, and organizations. *Sociology of Education* 74: 19–34. https://doi.org/10.2307/2673251.

Bruch, Elizabeth E., and M. E. J. Newman. 2018. Aspirational pursuit of mates in online dating markets. *Science Advances* 4(8):eaap9815. https://doi.org/10.1126/sciadv.aap9815.

Bruch, Elizabeth E., and M. E. J. Newman. 2019. Structure of online dating markets in U.S. cities. *Sociological Science* 6: 219–234. https://doi.org/10.15195/v6.a9.

Buss, David M., and David P. Schmitt. 2019. Mate preferences and their behavioral manifestations. *Annual Review of Psychology* 70 (1): 77–110. https://doi.org/10.1146/annurev-psych-010418-103408.

Buss, David M., Todd K. Shackelford, Lee A. Kirkpatrick, and Randy J. Larsen. 2001. A half century of mate preferences: The cultural evolution of values. *Journal of Marriage and the Family* 63 (2): 491–503. https://doi.org/10.1111/j.1741-3737.2001.00491.x.

Chadwick, Laura, and Gary Solon. 2002. Intergenerational income mobility among daughters. *American Economic Review* 92 (1): 335–344. https://doi.org/10.1257/000282802760015766.

Charles, Kerwin, Erik Hurst, and Alexandra Killewald. 2013. Marital sorting and parental wealth. *Demography* 50 (1): 51–70. https://doi.org/10.1007/s13524-012-0144-6.

Choi, Kate H., and Marta Tienda. 2017. Marriage-market constraints and mate-selection behavior: Racial, ethnic, and gender differences in intermarriage. *Journal of Marriage and Family* 79 (2): 301–317. https://doi.org/10.1111/jomf.12346.

Chudnovskaya, Margarita, and Ridhi Kashyap. 2020. Is the end of educational hypergamy the end of status hypergamy? Evidence from Sweden. *European Sociological Review* 36 (3): 351–365. https://doi.org/10.1093/esr/jcz065.

Davis, Kingsley. 1941. Intermarriage in caste societies. *American Anthropologist* 43 (3): 376–395.

DiMaggio, Paul, and John Mohr. 1985. Cultural capital, educational attainment, and marital selection. *American Journal of Sociology* 90 (6): 1231–1261. https://doi.org/10.1086/228209.

England, Paula. 2010. The gender revolution: Uneven and stalled. *Gender & Society* 24 (2): 149–166. https://doi.org/10.1177/0891243210361475.

England, Paula, and Su Li. 2006. Desegregation stalled: The changing gender composition of college majors, 1971–2002. *Gender & Society* 20 (5): 657–677. https://doi.org/10.1177/0891243206290753.

Esteve, Albert, Joan Garcia-Roman, and Inaki Permanyer. 2012. The gender-gap reversal in education and its effect on Union formation: The end of hypergamy? *Population and Development Review* 38 (3): 535–546. https://doi.org/10.1111/j.1728-4457.2012.00515.x.

Esteve, Albert, Christine R. Schwartz, Jan van Bavel, Inaki Permanyer, Martin Klesment, and Joan Garcia-Roman. 2016. The end of hypergamy: Global trends and implications. *Population and Development Review* 42 (4): 615–625.

Fu, Vincent Kang. 2010. Remarriage, delayed marriage, and Black/White intermarriage, 1968–1995. *Population and Development Review* 29 (5): 687–713. https://doi.org/10.1007/s11113-009-9168-z.

Goldscheider, Frances, Eva Bernhardt, and Trude Lappegård. 2015. The gender revolution: A framework for understanding changing family and demographic behavior. *Population and Development Review* 41 (2): 207–239. https://doi.org/10.1111/j.1728-4457.2015.00045.x.

Goldstein, Joshua R., and Catherine T. Kenney. 2001. Marriage delayed or marriage forgone? New cohort forecasts of first marriage for US women. *American Sociological Review* 66 (4): 506–519.

Goldstein, Joshua R., and Kristen Harknett. 2006. Parenting across racial and class lines: assortative mating patterns of new parents who are married, cohabiting, dating or no longer romantically involved. *Social Forces* 85 (1): 121–143.

Grow, André, and Jan van Bavel. 2015. Assortative mating and the reversal of gender inequality in education in Europe: An agent-based model ed. H. Reddy. *PLOS ONE* 10(6):e0127806. https://doi.org/10.1371/journal.pone.0127806.

Gullickson, Aaron, and Fu Vincent Kang. 2010. Comment: An endorsement of exchange theory in mate selection. *American Journal of Sociology* 115 (4): 1243–1251. https://doi.org/10.1086/649049.

Halpin, Brendan, and Tak Chan. 2003. Educational homogamy in Ireland and Britain: Trends and patterns. *British Journal of Sociology* 54 (4): 473–496. https://doi.org/10.1080/0007131032000143546.

Han, Hongyun. 2010. Trends in educational assortative marriage in China from 1970 to 2000. *Demographic Research* 22: 733–770. https://doi.org/10.4054/DemRes.2010.22.24.

Henz, Ursula, and Colin Mills. 2018. Social class origin and assortative mating in Britain, 1949–2010. *Sociology* 52 (6): 1217–1236. https://doi.org/10.1177/0038038517726479.

Hou, Feng, and John Myles. 2008. The changing role of education in the marriage market: Assortative marriage in Canada and the United States since the 1970s. *Canadian Journal of Sociology* 33 (2): 337–366.

Hout, Michael. 1982. The association between husbands' and wives' occupations in two-earner families. *American Journal of Sociology* 88 (2): 397–409.

Hu, Anning, and Zhenchao Qian. 2016. Does higher education expansion promote educational homogamy? Evidence from married couples of the post-80s generation in Shanghai, China. *Social Science Research* 60: 148–162. https://doi.org/10.1016/j.ssresearch.2016.05.001.
Hu, Yang. 2016. Marriage of matching doors: Marital sorting on parental background in China. *Demographic Research* 35: 557–580. https://doi.org/10.4054/DemRes.2016.35.20.
Kalmijn, Matthijs. 1991a. Shifting boundaries: Trends in religious and educational homogamy. *American Sociological Review* 56 (6): 786–800.
Kalmijn, Matthijs. 1991b. Status homogamy in the United States. *American Journal of Sociology* 97 (2): 496–523. https://doi.org/10.1086/229786.
Kalmijn, Matthijs. 1993. Trends in Black/White intermarriage. *Social Forces* 72 (1): 119–146. https://doi.org/10.2307/2580162.
Kalmijn, Matthijs. 1994. Assortative mating by cultural and economic occupational status. *American Journal of Sociology* 100 (2): 422–452.
Kalmijn, Matthijs. 1998. Intermarriage and homogamy: Causes, patterns, trends. *Annual Review of Sociology* 24: 395–421.
Kalmijn, Matthijs. 2013. The educational gradient in marriage: A comparison of 25 European Countries. *Demography* 50 (4): 1499–1520. https://doi.org/10.1007/s13524-013-0229-x.
Kalmijn, Matthijs, and Henk Flap. 2001. Assortative meeting and mating: Unintended consequences of organized settings for partner choices. *Social Forces* 79 (4): 1289–1312.
Lee, Soohyung. 2016. Effect of online dating on assortative mating: Evidence from South Korea. *Journal of Applied Econometrics* 31 (6): 1120–1139. https://doi.org/10.1002/jae.2480.
Lewis, Susan K., and Valerie K. Oppenheimer. 2000. Educational assortative mating across marriage markets: Non-hispanic whites in the United States. *Demography* 37 (1): 29–40. https://doi.org/10.2307/2648094.
Lichter, Daniel T., Diane K. McLaughlin, George Kephart, and David J. Landry. 1992. Race and the retreat from marriage: A shortage of marriageable men? *American Sociological Review* 57 (6): 781–799. https://doi.org/10.2307/2096123.
Lichter, Daniel T., Robert N. Anderson, and Mark D. Hayward. 1995. Marriage markets and marital choice. *Journal of Family Issues* 16 (4): 412–431. https://doi.org/10.1177/019251395016004001.
Lichter, Daniel T., Julie H. Carmalt, and Zhenchao Qian. 2011. Immigration and intermarriage among hispanics: Crossing racial and generational boundaries. *Sociological Forum* 26 (2): 241–264. https://doi.org/10.1111/j.1573-7861.2011.01239.x.
Lichter, Daniel T., Joseph P. Price, and Jeffrey M. Swigert. 2020. Mismatches in the marriage market. *Journal of Marriage and Family* 82 (2): 796–809. https://doi.org/10.1111/jomf.12603.
Lin, Ken-Hou., and Jennifer Lundquist. 2013. Mate selection in Cyberspace: The intersection of race, gender, and education. *American Journal of Sociology* 119 (1): 183–215. https://doi.org/10.1086/673129.
Lin, Zhiyong, Sonalde Desai, and Feinian Chen. 2020. The emergence of educational hypogamy in India. *Demography* 57 (4): 1215–1240. https://doi.org/10.1007/s13524-020-00888-2.
Mäenpää, Elina. 2015. Homogamy in educational level and parental social class in Finland: A log-linear analysis. *European Sociological Review* 31 (3): 253–267. https://doi.org/10.1093/esr/jcu088.
Mare, Robert D. 1991. Five decades of educational assortative mating. *American Sociological Review* 56 (1): 15–32.
McClendon, David. 2016. Religion, Marriage markets, and assortative mating in the United States: Religion, Marriage markets, assortative mating. *Journal of Marriage and Family* 78 (5):1399–1421. https://doi.org/10.1111/jomf.12353.
McClendon, David. 2018. Crossing boundaries: 'Some college', schools, and educational assortative mating: Some college, schools, assortative mating. *Journal of Marriage and Family* 80 (4): 812–825. https://doi.org/10.1111/jomf.12482.
Merton, Robert K. 1941. Intermarriage and the social structure: Fact and theory. *Psychiatry* 4 (3): 361–374.
Muhsam, H.V. 1974. The marriage squeeze. *Demography* 11 (2): 291–299.

Musick, Kelly, Jennie E. Brand, and Dwight Davis. 2012. Variation in the relationship between education and marriage: Marriage market mismatch? *Journal of Marriage and Family* 74 (1): 53–69. https://doi.org/10.1111/j.1741-3737.2011.00879.x.

Oppenheimer, Valerie Kincade. 1988. A theory of marriage timing. *American Journal of Sociology* 94 (3): 563–591.

Oppenheimer, Valerie Kincade. 1997. Women's employment and the gain to marriage: The specialization and trading model. *Annual Review of Sociology* 23: 431–453.

Parkin, Frank. 1971. *Class inequality and political order*. University of California Press.

Potarca, Gina. 2017. Does the Internet affect assortative mating? Evidence from the U.S. and Germany. *Social Science Research* 61: 278–297. https://doi.org/10.1016/j.ssresearch.2016.06.019.

Qian, Zhenchao. 1997. Breaking the racial barriers: Variations in interracial marriage between 1980 and 1990. *Demography* 34 (2): 263–276. https://doi.org/10.2307/2061704.

Qian, Yue. 2018. Educational assortative mating and income dynamics in couples: A longitudinal dyadic perspective. *Journal of Marriage and Family* 80 (3): 607–621. https://doi.org/10.1111/jomf.12470.

Qian, Yue, and Zhenchao Qian. 2017. Assortative mating by education and Hukou in Shanghai. *Chinese Sociological Review* 49 (3): 239–262. https://doi.org/10.1080/21620555.2017.1288066.

Qian, Zhenchao, and Daniel T. Lichter. 2007. Social boundaries and marital assimilation: Interpreting trends in racial and ethnic intermarriage. *American Sociological Review* 72 (1): 68–94.

Qian, Zhenchao, and Daniel T. Lichter. 2018. Marriage markets and intermarriage: Exchange in first marriages and remarriages. *Demography* 55 (3): 849–875. https://doi.org/10.1007/s13524-018-0671-x.

Raley, R. Kelly. 1996. A shortage of marriageable men? A note on the role of cohabitation in Black-White differences in marriage rates. *American Sociological Review* 61 (6): 973–983. https://doi.org/10.2307/2096303.

Rauscher, Emily. 2015. Effects of early U.S. compulsory schooling laws on educational assortative mating: The importance of context. *Demography* 52 (4): 1219–1242. https://doi.org/10.1007/s13524-015-0402-5.

Ravazzini, Laura, Ursina Kuhn, and Christian Suter. 2017. Do opposites attract? Educational assortative mating and dynamics of wage homogamy in Switzerland, 1992–2014. *Swiss Journal of Sociology* 43 (3): 567–586. https://doi.org/10.1515/sjs-2017-0028.

Raymo, James M., and Miho Iwasawa. 2005. Marriage market mismatches in Japan: An alternative view of the relationship between women's education and marriage. *American Sociological Review* 70 (5): 801–822. https://doi.org/10.1177/000312240507000504.

Raymo, James M., and Hyunjoon Park. 2020. Marriage decline in Korea: Changing composition of the domestic marriage market and growth in international marriage. *Demography* 57 (1): 171–194. https://doi.org/10.1007/s13524-019-00844-9.

Ridgeway, Cecilia L. 2001. Gender, status, and leadership. *Journal of Social Issues* 57 (4): 637–655. https://doi.org/10.1111/0022-4537.00233.

Ridgeway, Cecilia L., and Shelley J. Correll. 2004. Unpacking the gender system: A theoretical perspective on gender beliefs and social relations. *Gender & Society* 18 (4): 510–531. https://doi.org/10.1177/0891243204265269.

Ridgeway, Cecilia L. 2011. *Framed by Gender: How Gender inequality persists in the modern world*. Oxford University Press.

Rosenfeld, Michael J. 2005. A critique of exchange theory in mate selection. *American Journal of Sociology* 110 (5): 1284–1325.

Rosenfeld, Michael J. 2007. *The age of independence: Interracial Unions, Same-Sex Unions, and the changing American Family*. Harvard University Press.

Rosenfeld, Michael J. 2008. Racial, educational and religious endogamy in the United States: A comparative historical perspective. *Social Forces* 87 (1): 1–31.

Rosenfeld, Michael J. 2010. Still weak support for status-caste exchange: A reply to critics. *American Journal of Sociology* 115 (4): 1264–1276. https://doi.org/10.1086/649051.

Rosenfeld, Michael J., and Byung-Soo Kim. 2005. The independence of young adults and the rise of interracial and same-sex unions. *American Sociological Review* 70 (4): 541–62. https://doi.org/10.1177/000312240507000401.

Rosenfeld, Michael J., and Reuben J. Thomas. 2012. Searching for a mate: The rise of the internet as a social intermediary. *American Sociological Review* 77 (4): 523–547. https://doi.org/10.1177/0003122412448050.

Rosenfeld, Michael J., Reuben J. Thomas, and Sonia Hausen. 2019. Disintermediating your friends: How online dating in the United States displaces other ways of meeting. *Proceedings of the National Academy of Sciences* 116 (36): 17753–17758. https://doi.org/10.1073/pnas.1908630116.

Rytina, Steven, Peter Michael Blau, Terry Blum, and Joseph E. Schwartz. 1988. Inequality and intermarriage: A paradox of motive and constraint. *Social Forces* 66 (3): 645–675.

Schoen, Robert. 1983. Measuring the tightness of a marriage squeeze. *Demography* 20 (1): 61–78. https://doi.org/10.2307/2060901.

Schwartz, Christine R. 2010. Pathways to educational homogamy in marital and cohabiting unions. *Demography* 47 (3): 735–753.

Schwartz, Christine R. 2013. Trends and variation in assortative mating: Causes and consequences. *Annual Review of Sociology* 39 (1): 451–470. https://doi.org/10.1146/annurev-soc-071312-145544.

Schwartz, Christine R., and Robert D. Mare. 2005. Trends in educational assortative marriage from 1940 to 2003. *Demography* 42 (4): 621–646.

Schwartz, Christine R., and Robert D. Mare. 2012. The proximate determinants of educational homogamy: The effects of first marriage, marital dissolution, remarriage, and educational upgrading. *Demography* 49 (2): 629–650. https://doi.org/10.1007/s13524-012-0093-0.

Schwartz, Christine R., Zhen Zeng, and Yu Xie. 2016. Marrying up by marrying down: Status exchange between social origin and education in the United States. *Sociological Science* 3: 1003–1027. https://doi.org/10.15195/v3.a44.

Shafer, Kevin. 2013. Unique matching patterns in remarriage: Educational assortative mating among divorced men and women. *Journal of Family Issues* 34 (11): 1500–1535. https://doi.org/10.1177/0192513X12459020.

Shafer, Kevin, and Zhenchao Qian. 2010. Marriage timing and educational assortative mating. *Journal of Comparative Family Studies* 41 (5): 661–691.

Smits, Jeroen. 2003. Social closure among the higher educated: Trends in educational homogamy in 55 countries. *Social Science Research* 32 (2): 251–277. https://doi.org/10.1016/S0049-089X(02)00049-2.

Smits, Jeroen, and Hyunjoon Park. 2009. Five decades of educational assortative mating in 10 East Asian societies. *Social Forces* 88 (1): 227–255.

Smits, Jeroen, Wout Ultee, and Jan Lammers. 1998. Educational homogamy in 65 Countries: An explanation of differences in openness using country-level explanatory variables. *American Sociological Review* 63 (2): 264–285. https://doi.org/10.2307/2657327.

Sorensen, Annemette. 1994. Women, family and class. *Annual Review of Sociology* 20: 27–47.

South, Scott J. 1991. Sociodemographic differentials in mate selection preferences. *Journal of Marriage and the Family* 53 (4): 928. https://doi.org/10.2307/352998.

Sweeney, Megan M. 2002. Two decades of family change: The shifting economic foundations of marriage. *American Sociological Review* 67 (1): 132–147. https://doi.org/10.2307/3088937.

Thomas, Reuben J. 2019. Online exogamy reconsidered: Estimating the internet's effects on racial, educational, religious, political and age assortative mating. *Social Forces* 98 (3): 1257–1286. https://doi.org/10.1093/sf/soz060.

Tian, Felicia F., and Deborah S. Davis. 2019. Reinstating the family: Intergenerational influence on assortative mating in China. *Chinese Sociological Review* 51 (4): 337–364. https://doi.org/10.1080/21620555.2019.1632701.

Treiman, Donald J. 1970. Industrialization and social stratification. *Sociological Inquiry* 40 (2): 207–234.

Uunk, Wilfred J. G., Harry B. G. Ganzeboom, and Peter Róbert. 1996. Bivariate and multivariate scaled association models: An application to homogamy of social origin and education in Hungary between 1930 and 1979. *Quality & Quantity* 30: 323–343.

van Bavel, Jan. 2012. The reversal of the gender gap in education and female breadwinners in Europe ed. by W. Lutz, K. S. James, V. Skirbekk, and J. van Bavel. *Vienna Yearbook of Population Research* 10: 127–154. https://doi.org/10.1553/populationyearbook2012s127.

van Bavel, Jan, Christine R. Schwartz, and Albert Esteve. 2018. The reversal of the gender gap in education and its consequences for family life. *Annual Review of Sociology* 44: 341–360. https://doi.org/10.1146/annurev-soc-073117-041215.

Wang, Hongyu, Grace Kao, and Kara Joyner. 2006. Stability of interracial and intraracial romantic relationships among adolescents. *Social Science Research* 35 (2): 435–453. https://doi.org/10.1016/j.ssresearch.2004.10.001.

Western, Bruce. 2006. *Punishment and inequality in America*. Russell Sage Foundation.

Wong, Raymond Sin-Kwok. 2003. To see or not to see: Another look at research on temporal trends and cross-national differences in educational homogamy. *Taiwanese Journal of Sociology* 31: 45–88.

Xie, Yu, Siwei Cheng, and Xiang Zhou. 2015. Assortative mating without assortative preference. *Proceedings of the National Academy of Sciences* 112 (19): 5974–5978. https://doi.org/10.1073/pnas.1504811112.

Chapter 3
How We Evaluate Assortative Mating

3.1 Introduction

In this chapter, we briefly review methodological issues in research on assortative mating. The motivation for this chapter is recognition that our understanding of trends or consequences of assortative mating is shaped by the methods we choose. It is thus critical to discuss the (dis)advantages of using particular methods as well as how the rationale for the use of particular methods and tools depends upon our specific research question(s). For simplicity, we limit our discussion to a small number of well-established methods. In measuring patterns and trends, the standard approach has been log-linear models, especially when trying to estimate the likelihood or the odds of educational homogamy. The harmonic mean model is an alternative approach well suited to the description of marriage rates for specific pairing patterns and estimation of the impacts of changing partner availability in the marriage market. In studies of the consequences of assortative mating, counterfactual tools have often been employed to estimate the impact of trends in assortative mating on certain outcomes, most notably income inequality. We also review a regression approach that is suitable to use when our estimand is an individual-level outcome.

3.2 Measuring Patterns and Trends

3.2.1 Log-Linear Models

Studies of educational assortative mating are typically interested in aggregate union formation patterns like homogamy, hypergamy, or hypogamy, net of marginal distributions. This is because the prevalence of different pairings (and change therein) reflects the educational distribution (e.g., women's relative increase in higher education) as well as the "true" likelihood of homogamy, which scholars interpret as an

F. Uchikoshi and J. M. Raymo, *Educational Assortative Mating in Japan*,
Population Studies of Japan, https://doi.org/10.1007/978-981-16-3713-1_3

indicator of preferences (Schwartz 2013).[1] By controlling for compositional changes in men's and women's educational attainment, we can focus on changes in the relative odds of homogamy (or any other pairing type of interest) conditional on educational attainment distributions. To estimate these odds ratios and their changes over time, prior studies have typically applied log-linear (and log-multiplicative models) to contingency tables.

Suppose that we have a three-way table of husband's educational attainment $H(i = 1, \ldots, 5)$, wife's educational attainment $W(j = 1, \ldots, 5)$, and marriage cohort $C(k = 1, \ldots, 3)$. A common baseline model is the conditional independence model, which assumes independence of wives' and husbands' educational achievement and is typically expressed as:

$$\ln F_{ijk} = \lambda + \lambda_i^H + \lambda_j^W + \lambda_k^C + \lambda_{ik}^{HC} + \lambda_{jk}^{WC}$$

where F_{ijk} is the expected frequency of marriages in the (i, j, k) cell, i.e., the number of marriages between men of education level i and women of education level j in marriage cohort k. For example, F_{543} indicates the frequency of couples in which the husbands with the highest educational attainment, the wives with the second highest, and the marriage occurred in the latest period in the analysis. In this model, educational pairing is assumed to be random, net of marginal distributions. Extending this baseline model by adding additional parameters (i.e., structure) and comparing model fit allows researchers to choose a pattern that best represents the data or to evaluate theoretically derived hypotheses about the nature of spouse pairing (and differences across time or space). For example, to test whether the likelihood of educational homogamy is significantly larger than other types of pairing, one could use the quasi-independence model of homogamy:

$$\begin{aligned} \ln F_{ijk} &= \lambda + \lambda_i^H + \lambda_j^W + \lambda_k^C + \lambda_{ik}^{HC} + \lambda_{jk}^{WC} + \delta_{ij}^{HW}, \\ \text{where } \delta_{ij}^{HW} &= 0 \text{ for i} \neq \text{j} \end{aligned}$$

This model estimates the likelihood, or more precisely the log odds, of homogamy by explicitly parameterizing educationally homogamous marriages (δ_{ij}^{HW}). More generally, we can use design matrices (examples are provided in Chap. 5) that represent more complex pairing patterns, and allow for flexible evaluation of any theoretically motivated pattern of spouse pairing.[2]

---

[1] Previous studies frequently associated the likelihood of specific pairings, net of composition (and compositional change), as a reflection of preferences (Choi and Mare 2012; Hou and Myles 2008; Schwartz and Mare 2005). As Schwartz (2013, footnote 6) noted, however, estimated parameters from log-linear models reflect not only preferences, but also opportunities, and other structural forces.

[2] It should be noted here that some scholars question the use of log-linear models to understand individual preferences given that estimated pairing parameters reflect, by definition, the preferences of both partners (Schwartz 2013, footnote 6). In an effort to address this problem, Logan et al. (2008) proposed the two-sided logit model, an approach that allows for estimation of pairing preferences for each partner, but this has not been widely utilized.

3.2.2 Harmonic Mean Model Approach

One critique of the log-linear model is that includes no information about the population at the risk of marriage, i.e., singles. By focusing on already married couples, conventional log-linear models are unable to distinguish preferences from opportunity structure (Logan et al. 2008). This drawback is critical if one's interest is to examine the role of marriage market composition with respect to characteristics of interest such as educational attainment. To examine how changes in the pool of potential partners influence patterns of assortative mating, several studies have used the harmonic mean models of marriage (Fukuda et al. 2020; Qian and Preston 1993; Raymo and Iwasawa 2005; Raymo and Park 2020). By incorporating the population at risk of first marriage (rather than limiting the focus to married couples), these models allow for a straightforward evaluation of the role of both marriage market composition and pairing propensities in determining marriage rates (the measure of ultimate interest). The model, first proposed by Schoen (1988), can be written as:

$$N_{ijkl}^{t} = \left(\frac{N_{ijkl}^{t}}{F_{ik}^{t}} + \frac{N_{ijkl}^{t}}{M_{jl}^{t}}\right) \times \frac{M_{jl}^{t} F_{ik}^{t}}{F_{ik}^{t} + M_{jl}^{t}} = \alpha_{ijkl}^{t} \times \frac{M_{jl}^{t} F_{ik}^{t}}{F_{ik}^{t} + M_{jl}^{t}}$$

where N_{ijkl}^{t} is the number of first marriages of women of age (i) and educational attainment (k) paired with husbands of age (j) with educational attainment (l) in period (t). Following Schoen (1988), the likelihood of the specific pairing of interest, net of marriage market composition, is called the "force of attraction" (α_{ijkl}^{t}). According to Qian and Preston (1993: 483), the force of attraction reflects the rate of encounters between men and women in the marriage market and the proportion of such encounters that lead to marriage.

By dividing the number of marriages by the male or female population at risk, we can calculate first marriage rates. For example, the marriage rate for women can be expressed as:

$$m_{ijkl}^{f,t} = \frac{N_{ijkl}^{t}}{F_{ik}^{t}} = \left(\frac{N_{ijkl}^{t}}{F_{ik}^{t}} + \frac{N_{ijkl}^{t}}{M_{jl}^{t}}\right) \times \frac{M_{jl}^{t}}{F_{ik}^{t} + M_{jl}^{t}} = \alpha_{ijkl}^{t} \times \frac{M_{jl}^{t}}{F_{ik}^{t} + M_{jl}^{t}}$$

where $m_{ijkl}^{f,t}$ is the first marriage rate for women (f) of age (i) and educational attainment (k) paired with husbands age (j) with educational attainment (l) at period (t). The marriage rate is calculated as the number of marriage (N_{ijkl}^{t}) divided by the number of women of these ages (i) and educational attainment (k) exposed to the risk of marriage at time (t). This rate is thus the product of the force of attraction and the availability of single men and women who are at risk of marriage $\frac{M_{jl}^{t}}{F_{ik}^{t}+M_{jl}^{t}}$, which has been called an "availability ratio" (Qian and Preston 1993; Raymo and Iwasawa 2005).

3.3 Evaluating Impacts

3.3.1 *Counterfactual Approach*

The use of counterfactual analyses allows us to quantify the extent to which changes in marriage market composition with respect to educational attainment have contributed to changing marriage rates. This approach also allows us to evaluate the extent to which changes in pairing propensities may have exacerbated or mitigated the impact of changing market composition on marriage rates. To evaluate the contribution of changes in marriage market composition (i.e., the availability ratio in the harmonic mean models of marriage described above) to changes in marriage rates, we can calculate counterfactual marriage rates in which the availability ratio is held constant at its initial value (or some earlier value, more generally).[3] For instance, counterfactual marriage rates for 2010 using marriage market composition observed in 2000 are calculated as:

$$m_{ijkl}^{f,\,2010*} = \alpha_{ijkl}^{2010} \times \frac{M_{jl}^{2000}}{F_{ik}^{2000} + M_{jl}^{2000}}$$

By describing both observed and counterfactual marriage rates, we are able to assess the relative importance of changes in marriage market composition and changes in the propensity for specific pairings. In other words, these comparisons allow us to answer the following counterfactual question "what would age-education-specific marriage rates be in 2010 if marriage market composition had not changed since 2000?"

A counterfactual approach is also useful for examining the implications of patterns of assortative mating on income inequality via the use of decomposable measures such as the Theil index (Boertien and Permanyer 2019; Breen and Anderson 2012; Breen and Salazar 2010, 2011; Herzberg-Druker and Steir 2019; Hu and Qian 2015; Zagel and Breen 2019). To illustrate its relevance for research on assortative mating, we begin with a basic definition of the Theil index as it is used in research on income inequality:

$$T = \frac{1}{n}\sum_{i=1}^{n}\frac{x_i}{\bar{x}}\ln\frac{x_i}{\bar{x}}$$

where n is the number of households, x_i is the earnings of household i, $\bar{x}$ is the average earnings across households. The lowest value is 0, meaning that household

[3] Importantly, the same kind of counterfactual approach is possible with log-linear models. Although we are not aware of the application of the counterfactual approach for questions about homogamy, Breen (2010) used this approach to examine the impact of educational expansion on social mobility by distinguishing the contribution of the marginal distributions from that of mobility coefficients.

income is equally distributed across households, while the highest value is 1, which indicates perfect inequality.

One methodological advantage of the Theil index compared with other inequality measures is the ability to decompose the index into within-group inequality (T_w) and between-group inequality (T_b). Applying this index to income inequality and patterns of assortative mating, T for the *j*th educational pairing group (T_j) can be written as follows:

$$T_j = \frac{1}{n_j} \sum_{i=1}^{n_j} \frac{x_{ij}}{\bar{x}_j} \ln \frac{x_{ij}}{\bar{x}_j}$$

where n_i is the number of households within the jth group, x_{ij} is the household earnings of household i within the jth group, $\overline{x_j}$ is the mean of the household earnings within group j. Using the group-specific Theil index, we can decompose the overall Theil index into a between-group component (T_b) and a within-group component (T_w) as follows.

$$T = \sum_j p_j \frac{\bar{x}_j}{\bar{x}} \ln \frac{\bar{x}_j}{\bar{x}} + \sum_j p_j \frac{\bar{x}_j}{\bar{x}} T_j = \left(\sum_j p_j \frac{\bar{x}_j}{\sum_j p_j \bar{x}_j} \ln \frac{\bar{x}_j}{\sum_j p_j \bar{x}_j} \right) + \left(\sum_j p_j \frac{\bar{x}_j}{\sum_j p_j \bar{x}_j} T_j \right) = T_b + T_w$$

The between-group component represents inequality across groups measured by the group-mean using the share of educational pairing type p_j as a weight. The within-group component is the weighted average of the Theil index of each group (T_j), where the weights are the share of each educational pairing group.

As the equation above shows, the overall Theil index is comprised of three components, the group-specific level of income inequality (T_j), the average income within the group ($\bar{x}_j$), and the prevalence of different educational pairings (p_j). This property facilitates a range of interesting counterfactual analyses. For example, in analyses of how changing patterns of assortative mating are related to change in income inequality, scholars typically compare the observed level of income inequality at time t with a counterfactual level of inequality at time t calculated by holding patterns of of educational pairing (p_{1j}) at their previous (t-1) values. In a two-period scenario, the counterfactual Theil index (T_{cf}) for this analysis would be calculated as follows.

$$T_{cf} = \sum_j p_{2j} \frac{\bar{x}_{1j}}{\sum_j p_{2j} \bar{x}_{1j}} \ln \frac{\bar{x}_{1j}}{\sum_j p_{2j} \bar{x}_{1j}} + \sum_j p_j \frac{\bar{x}_{1j}}{\sum_j p_{2j} \bar{x}_{1j}} T_{1j}.$$

3.3.2 Regression Approach

Analyses of the effects of assortative mating on individual (and family) life-course outcomes, such as fertility or divorce, have typically estimated regression models that include interactions between husbands' and wives' educational attainment. Eeckhaut et al. (2013) note that the interaction effects of couples' education levels have been specified in two different ways. One is the absolute difference approach, which focuses on the difference in years of schooling between wives and husbands. The other is the categorical difference approach which focuses on levels of attainment to classify couples' educational pairing as homogamous, hypergamous, or hypogamous. Note that regression models are also frequently used to study assortative mating at the individual level, with multinomial logistic regression models or competing risk hazard models for different pairing outcomes specified as a function of characteristics of interest (Hou and Myles 2013; Lewis and Oppenheimer 2000; Raymo and Iwasawa 2008; Schwartz and Graf 2009; Uchikoshi 2018).

Despite its wide use and simplicity, there are several drawbacks to a regression-based approach to the study of assortative mating. The two different approaches to measuring educational homogamy have advantages and disadvantages. The absolute difference approach uses fewer degrees of freedom and generates coefficients that are easy to interpret, but it is often criticized for ignoring the qualitative differences between different levels of education (degrees). The categorical difference approach reflects qualitative differences in levels of educational attainment, but does not estimate "main effects" of educational attainment. For example, only defining three variables, homogamy, hypergamy, hypogamy, ignores the main effects from each spouse's education. Even when including the wife's and husband's educational attainment, in addition to the interaction terms, it is not straightforward to interpret the interaction coefficients (Eeckhaut et al. 2013). This approach also suffers from the linear dependency problem because the difference between husband's education (x_1) and wife's education (x_2) is the interaction term (x_1-x_2). Alternative approaches that solve problems with interaction effects include the diagonal reference model (Sobel 1981), where the reference group is homogamous couples (in two-way cross tabulations of frequencies F_{ij} with n rows and columns, they are located on the diagonal cells, F_{kk} where k = 1,2,..n) and we estimate the effects of pairing patterns from a weighted average of two diagonal cells (e.g., to estimate F_{ij}, we use the information about F_{ii} and F_{jj}, where i,j = 1,2,…n, respectively). But these methods are rarely used in studies of assortative mating (see Eeckhaut et al. (2014) and Eeckhaut and Stanfors (2021) for exceptions).

Another issue is the difficulty of understanding counterfactuals at the individual level (Miller 2020). This reflects the fact that the three broad categories of spouse pairing (homogamy, hypergamy, and hypogamy) are not always feasible outcomes. For example, women with the highest level of education (typically university or more) have no opportunity to marry up while women in the lowest educational category cannot marry down. To put it differently, hypergamy is not an available counterfactual for the highest-educated women while hypogamy is not an available counterfactual for the least educated. As such, we need to think hard about hypothetical counterfactuals when estimating the effects, especially *causal* effects, of assortative mating on individual-level outcomes as long as we rely on the potential outcome framework to estimate causal effects, which posits that the causal effect is estimable by taking the difference between the outcome with treatment and the outcome without treatment. For example, if our theoretical interest is the effect of educational hypogamy on some outcome of interest (e.g., marital satisfaction), this may require us to exclude those in the lowest educational category from our analytical sample. On this point, we believe the regression approach using OLS or logistic regression models would allow more flexible modeling rather than using a diagonal reference model that assumes a symmetrical relationship between spouses' subgroups.

Finally, when we think about how patterns of spouse pairing affect outcomes of interest, it is critical to distinguish between additive mechanisms and interactive mechanisms. The former refers to the socioeconomic resources that husbands' and wives' educational attainment independently bring to the household. For example, it is clear that the socioeconomic status of each parent has independent influences on children's outcomes (Beller 2009). An increase in homogamy among the highly educated should contribute to growing household level inequality via this mechanism. The latter mechanism refers to how specific combinations of husbands' and wives' education may mitigate or exacerbate the additive influences, a central focus in research on spouse pairing and individual or family outcomes (Kalmijn 1998). To give one example, research showing that agreement on parenting style has a positive influence on children's cognitive outcomes (Martin et al. 2007) and research showing that parenting style differs by parents' socioeconomic status (Chan and Koo 2011; Doepke and Zilibotti 2019; Lareau 2011; Kalil and Ryan 2020; Lundberg et al. 2016; Weininger et al. 2015) suggests that patterns of educational pairing should be related to children's cognition net of each parent's educational attainment. This distinction between additive and interactive effects of spouse pairing patterns applies to other outcomes, including several that we consider in subsequent chapters (fertility, divorce, and the spousal division of labor).

References

Beller, Emily. 2009. Bringing intergenerational social mobility research into the twenty-first century: Why mothers matter. *American Sociological Review* 74 (4): 507–528.

Boertien, Diederik, and Iñaki Permanyer. 2019. Educational assortative mating as a determinant of changing household income inequality: A 21-country study. *European Sociological Review* 35 (4): 522–537. https://doi.org/10.1093/esr/jcz013.

Breen, Richard. 2010. Educational expansion and social mobility in the 20th century. *Social Forces* 89 (2): 365–388. https://doi.org/10.1353/sof.2010.0076.

Breen, Richard, and Signe Hald Andersen. 2012. Educational assortative mating and income inequality in Denmark. *Demography* 49 (3): 867–887. https://doi.org/10.1007/s13524-012-0111-2.

Breen, Richard, and Leire Salazar. 2010. Has increased women's educational attainment led to greater earnings inequality in the United Kingdom? A multivariate decomposition analysis. *European Sociological Review* 26 (2): 143–157. https://doi.org/10.1093/esr/jcp011.

Breen, Richard, and Leire Salazar. 2011. Educational assortative mating and earnings inequality in the United States. *American Journal of Sociology* 117 (3): 808–843. https://doi.org/10.1086/661778.

Chan, Tak Wing, and Anita Koo. 2011. Parenting style and youth outcomes in the UK. *European Sociological Review* 27 (3): 385–399. https://doi.org/10.1093/esr/jcq013.

Choi, Kate H., and Robert D. Mare. 2012. International migration and educational assortative mating in Mexico and the United States. *Demography* 49 (2): 449–476. https://doi.org/10.1007/s13524-012-0095-y.

Doepke, Matthias, and Fabrizio Zilibotti. 2019. *Love, money and parenting: How economics explains the way we raise our kids*. Princeton University Press.

Eeckhaut, Mieke C. W., and Maria A. Stanfors. 2021. Educational assortative mating, gender equality, and income differentiation across Europe: A simulation study. *Acta Sociologica* 64(1): 48-69. https://doi.org/10.1177/0001699319877925.

Eeckhaut, Mieke C. W., Bart Van de Putte, Jan R. M. Gerris, and Ad.A. Vermulst. 2013. Analysing the effect of educational differences between partners: A methodological/theoretical comparison. *European Sociological Review* 29 (1): 60–73. https://doi.org/10.1093/esr/jcr040.

Eeckhaut, Mieke C. W., Maria A. Stanfors, and Bart Van de Putte. 2014. Educational heterogamy and the division of paid labour in the family: A comparison of present-day Belgium and Sweden. *European Sociological Review* 30 (1): 64–75. https://doi.org/10.1093/esr/jct022.

Fukuda, Setsuya, James M. Raymo, and Shohei Yoda. 2020. Revisiting the educational gradient in marriage in Japan. *Journal of Marriage and Family* 82 (4): 1378–1396. https://doi.org/10.1111/jomf.12648.

Herzberg-Druker, Efrat, and Haya Stier. 2019. Family matters: The contribution of households' educational and employment composition to income inequality. *Social Science Research* 82: 221–239. https://doi.org/10.1016/j.ssresearch.2019.04.012.

Hou, Feng, and John Myles. 2008. The changing role of education in the marriage market: Assortative marriage in Canada and the United States Since the 1970s. *Canadian Journal of Sociology* 33 (2): 337–366.

Hou, Feng, and John Myles. 2013. Interracial marriage and status-caste exchange in Canada and the United States. *Ethnic and Racial Studies* 36 (1): 75–96. https://doi.org/10.1080/01419870.2011.634505.

Hu, Anning, and Zhenchao Qian. 2015. Educational homogamy and earnings inequality of married couples: Urban China, 1988–2007. *Research in Social Stratification and Mobility* 40: 1–15. https://doi.org/10.1016/j.rssm.2015.01.004.

Kalil, Ariel, and Rebecca Ryan. 2020. Parenting practices and socioeconomic gaps in childhood outcomes. *The Future of Children* 30: 29–54. https://doi.org/10.1353/foc.2020.0004.

Kalmijn, Matthijs. 1998. Intermarriage and homogamy: Causes, patterns, trends. *Annual Review of Sociology* 24: 395–421.

Lareau, Annette. 2011. *Unequal childhoods: Class, race, and family life*, 2nd ed. University of California Press.

Lewis, Susan K., and Valerie K. Oppenheimer. 2000. Educational assortative mating across marriage markets: Non-hispanic whites in the United States. *Demography* 37 (1): 29–40. https://doi.org/10.2307/2648094.

Logan, John Allen, Peter D. Hoff, and Michael A. Newton. 2008. Two-sided estimation of mate preferences for similarities in age, education, and religion. *Journal of the American Statistical Association* 103 (482): 559–569. https://doi.org/10.1198/016214507000000996.

Lundberg, Shelly, Robert A. Pollak, and Jenna Stearns. 2016. Family inequality: Diverging patterns in marriage, cohabitation, and childbearing. *Journal of Economic Perspectives* 30 (2): 79–102. https://doi.org/10.1257/jep.30.2.79.

Martin, Anne, Rebecca M. Ryan, and Jeanne Brooks-Gunn. 2007. The joint influence of mother and father parenting on child cognitive outcomes at age 5. *Early Childhood Research Quarterly* 22(4):423–39. https://doi.org/10.1016/j.ecresq.2007.07.001.

Miller, Rhiannon N. 2020. Educational assortative mating and time use in the home. *Social Science Research* 90:102440. https://doi.org/10.1016/j.ssresearch.2020.102440.

Qian, Zhenchao, and Samuel H. Preston. 1993. Changes in American marriage, 1972 to 1987: Availability and forces of attraction by age and education. *American Sociological Review* 58 (4): 482–495.

Raymo, James M., and Miho Iwasawa. 2005. Marriage market mismatches in Japan: An alternative view of the relationship between women's education and marriage. *American Sociological Review* 70 (5): 801–822. https://doi.org/10.1177/000312240507000504.

Raymo, James M., and Miho Iwasawa. 2008. Bridal pregnancy and spouse pairing patterns in Japan. *Journal of Marriage and Family* 70 (4): 847–860. https://doi.org/10.1111/j.1741-3737.2008.00531.x.

Raymo, James M., and Hyunjoon Park. 2020. Marriage decline in Korea: Changing composition of the domestic marriage market and growth in international marriage. *Demography* 57 (1): 171–194. https://doi.org/10.1007/s13524-019-00844-9.

Schoen, Robert. 1988. *Modeling multigroup populations*. Plenum Press.

Schwartz, Christine R. 2013. Trends and variation in assortative mating: Causes and consequences. *Annual Review of Sociology* 39 (1): 451–470. https://doi.org/10.1146/annurev-soc-071312-145544.

Schwartz, Christine R., and Nikki L. Graf. 2009. Assortative matching among same-sex and different-sex couples in the United States, 1990–2000. *Demographic Research* 21: 843–878. https://doi.org/10.4054/DemRes.2009.21.28.

Schwartz, Christine R., and Robert D. Mare. 2005. Trends in educational assortative marriage from 1940 to 2003. *Demography* 42 (4): 621–646.

Sobel, Michael E. 1981. Diagonal mobility models: A substantively motivated class of designs for the analysis of mobility effects. *American Sociological Review* 46 (6): 893–906.

Uchikoshi, Fumiya. 2018. Assortative mating in the age of marriage decline. *Sociological Theory and Methods* 33 (1): 15–31 (in Japanese). https://doi.org/10.11218/ojjams.33.15.

Weininger, Elliot B., Annette Lareau, and Dalton Conley. 2015. What money doesn't buy: Class resources and children's participation in organized extracurricular activities. *Social Forces* 94 (2): 479–503. https://doi.org/10.1093/sf/sov071.

Zagel, Hannah, and Richard Breen. 2019. Family demography and income inequality in West Germany and the United States. *Acta Sociologica* 62 (2): 174–192. https://doi.org/10.1177/0001699318759404.

Chapter 4
The Japanese Context

As discussed earlier, changing attitudes and expectations regarding gender are seen as one of the most salient factors for understanding recent trends in educational assortative mating and their consequences. In this respect, the Japanese case provides an important opportunity to better understand how trends in educational assortative mating in societies characterized by persistent gender inequality may or may not resemble those documented in more gender-egalitarian societies. In gender-inegalitarian Japan, longstanding preferences on the part of both men and women for educational hypergamy and the associated breadwinner–homemaker division of labor are thought to provide highly educated women with strong incentives to remain focused on their career and avoid marriage. However, this conventional wisdom regarding gender context, assortative mating, and non-marriage may be of declining relevance in recent years as men's economic prospects gradually decline, and both men and women increasingly prefer dual-earner marriages. In this evolving context, some studies have reported an increase in women's educational hypogamy. These trends may be enhanced by the shifting composition of the marriage market due to the relatively rapid growth of women in higher education. Our goal in this chapter is to provide a detailed summary of relevant social and economic changes in recent decades.

4.1 Introduction

As earlier studies have noted, patterns and trends in educational assortative mating vary across countries (Schwartz 2013), suggesting the relevance of institutional forces. Building upon the international literature on assortative mating, this section discusses several important features of the Japanese context relevant to our understanding of educational pairing patterns and their consequences. We pay particular attention to the meaning of marriage, the gender difference in status attainment, and changing patterns of family formation. Understanding these contextual features

F. Uchikoshi and J. M. Raymo, *Educational Assortative Mating in Japan*,
Population Studies of Japan, https://doi.org/10.1007/978-981-16-3713-1_4

requires attention to the gendered nature of social and economic institutions. For example, key features of the labor market, including seniority(age)-based promotion, expectations of lifetime employment, dedication and commitment to one's employer, and long work hours are all predicated on the assumption of families comprised of one spouse working full-time in the breadwinner role and the other spouse responsible for household work and caring for a spouse, children, and occasionally coresident parents(-in-law) (Brinton 1993; Estevez-Abe 2008; Osawa 1993; Tsutsui 2020; Yamaguchi 2019). This assumption is also reflected in the career track hiring system, in which one track offers the possibility of promotion (career track, *sōgōshoku*) and the other does not (non-career track, *ippanshoku*). Almost all male employees pursue the former track while a large proportion of women choose to pursue the latter track (Brinton 2007; Ishida et al. 2002; Nemoto 2016; Ogasawara 1998). These gender-segregated, "either-or" employment tracks limit career opportunities for women, especially those with higher levels of education and skills and more motivation to work after marriage and childbearing (Brinton 2001). Women's disadvantage in the labor market means that marriage has been long considered a primary mechanism for their social mobility in Japan (Shirahase 2001; Yamada 1996). This gender asymmetry results in a high prevalence of female hypergamy with respect to both age and educational attainment (Brinton et al. 2021; Fujihara and Uchikoshi 2019; Fujimi 2009; Raymo and Iwasawa 2005). At the same time, however, women's increasing participation in higher education makes it necessary to reconsider the connection between gender inequality and spouse pairing patterns. Of particular importance is the fact that the preference for women to marry up with respect to educational attainment is increasingly constrained by the fact that the relative supply of highly educated men in the marriage market has shrunk. As previous studies have argued, this marriage market mismatch has played a role in explaining declining marriage rates in Japan (Raymo and Iwasawa 2005).

Focusing on distinctive gender differences in educational attainment, evidence of women's preference for educational hypergamy, as well as recent sociodemographic changes, this chapter provides a concise, but comprehensive, contextual overview designed to help us to make sense of patterns of educational assortative mating in Japan.

4.2 The Meaning of Marriage for Men and Women in Japan

Throughout the post-World War II period, the socioeconomic status of women has been determined largely by that of their husbands (Shirahase 2001; Yasuda 1971). This gender asymmetry in status attainment is, of course, not unique to Japan. The relationship between spouse selection and women's social mobility in the United States and elsewhere has been studied extensively (Chase 1975; Cheng and Dai 1995; Erikson and Goldthorpe 1992; Glenn et al. 1974; Heath 1981; Portocarero

1985; Rossi 1965; Tyree and Treas 1974). Describing gender differences in the determination of social status in the United States, Rossi (1965: 1198) stated: "What a man 'does' defines his status, but whom she marries defines a woman's." While this claim is surely outdated in many countries where women's educational and occupational status are thought to be increasingly important in determining their partners' social status (Blossfeld and Timm 2003; Oppenheimer 1988; Schwartz and Mare 2005; van Bavel et al. 2018), it certainly maintains validity in contemporary Japan, where labor market dynamics have perpetuated gender asymmetry in the determination of social status. The Japanese labor market has long been characterized by high returns to continuous job tenure and experience (Clark and Ogawa 1992; Hashimoto and Raisian 1985; Kalleberg and Lincoln 1988; Mincer and Higuchi 1988; Yamada and Kawaguchi 2015), although the practice of long-term employment has declined somewhat during the prolonged economic downturn following the collapse of the bubble economy in 1991 (Kawaguchi and Ueno 2013). In this labor market context, achieving economic independence or social status through paid employment has been, and continues to be, extremely difficult for women who temporarily leave the labor force to raise children (Brinton 1988, 1991, 2001; Nemoto 2016; Ogawa and Clark 1995; Osawa 1993; Yu 2006).

Either motivated by a personal desire to concentrate efforts on homemaking and child care (Takeuchi 2004; Yamada 1996; Zhang 2012) or compelled by social pressures and/or the structural difficulties of combining work and child care (Iwasawa 1999; Ogasawara 1998; Yu 2009; Zhou 2015), labor force exit around marriage or first childbirth has long been the modal experience for women in Japan. The prevalence of this pattern of temporary or permanent exit from the labor force is the source of the so-called M-shaped age profile of female employment (Brinton 2001) and explains why women's educational attainment is not strongly associated with labor force participation (Brinton and Lee 2001). Yamada (1996) argues that the large proportion of single women who express a desire to leave the labor force to focus on domestic work following marriage or childbirth is a manifestation of women's desire to be reborn through marriage (*umarekawari shikō*). Yamada further argues that the financial ability to concentrate (at least temporarily) on domestic concerns is a minimum criterion for a "good" rebirth, pointing again to gender asymmetry in the determination of socioeconomic status and the associated role of gender-asymmetric spouse selection criteria. Specifically, men, whose socioeconomic status is unaffected by marriage, place a higher priority on appearance and domestic skills than on earnings power when evaluating potential spouses (IPSS 2017). Seeking a wife who will not interfere with their career goals, men have preferred to marry women lower in status than themselves. Women, whose socioeconomic status is determined by whom they marry, have attached more value to economic characteristics. More specifically, women seek husbands whose occupation and earnings potential promise to provide them with a standard of living higher than (or at least equal to) that they could achieve by not marrying. A "good" marriage for women has, therefore, been one that is status hypergamy, or at least status homogamy.

To summarize, Yamada (1996) posited the following gender differences in spouse selection criteria. Men prefer to marry women whose educational attainment, occupational status, firm size, and age are lower than their own. Women prefer to marry men who are older than themselves and whose educational attainment, occupational status, and firm size exceed both their own and that of their fathers (Yamada 1996: 55). The limited emphasis on women's occupation and earnings potential corresponds to the weak relationship between married women's educational attainment and labor force participation (Brinton and Lee 2001: 134). The very low prevalence of childless marriages (Hara 2008) suggests that low levels of labor force participation among highly educated women following marriage and childbirth reflect an emphasis on the role of women's human capital in supporting children's development and educational success (Hirao 2001; Holloway 2010; Tsuya and Choe 2004; Yu 2009). Demographers have thus suggested that the combination of marriage, childbearing, and childrearing characterized by mothers' intensive investment in children's success, and care for elderly family members forms a "marriage package" (Bumpass et al. 2009) that is a distinguishing feature of the East Asian family context (Raymo et al. 2015).

4.3 Changing Education and Labor Market Opportunities

Focusing on the relationship between improvements in women's socioeconomic status, marriage market composition, assortative mating, and age at marriage, Yamada's explanation of changes in Japanese marriage timing closely resembles Oppenheimer's (1988) search-theoretic explanation of changes in marriage timing in the United States. The central premise of Oppenheimer's theory is that later age at marriage increases the difficulty of mating homogamously. Assuming that women have standards for a "minimally acceptable match" and that these standards are relatively inflexible, she argues that homogamy becomes increasingly difficult as relative improvements in women's economic status reduce the supply of economically attractive men. At the same time, absolute improvements in women's financial status facilitate longer searches for more attractive men. In sum, women's age at marriage increases as it becomes numerically more difficult to locate an acceptable spouse and financially less difficult to spend more time searching for one. Because higher status women have more restrictive spouse selection criteria as well as greater financial resources, delays in marriage caused by prolonged spouse search will be particularly pronounced for this group.

It is easy to see the relevance of Oppenheimer's theoretical perspective to Japan, where large gender differences in educational and occupational status have long guaranteed the prevalence of desirable matches (i.e., female status hypergamy/male status hypogamy). An extension of Oppenheimer's spouse search theory to the Japanese marriage market leads to the so-called marriage market mismatch hypothesis, which posits that these relative improvements in women's educational and occupational opportunities have made the process of locating an acceptable spouse more difficult and time-consuming by reducing the numerical opportunities for desirable pairings.

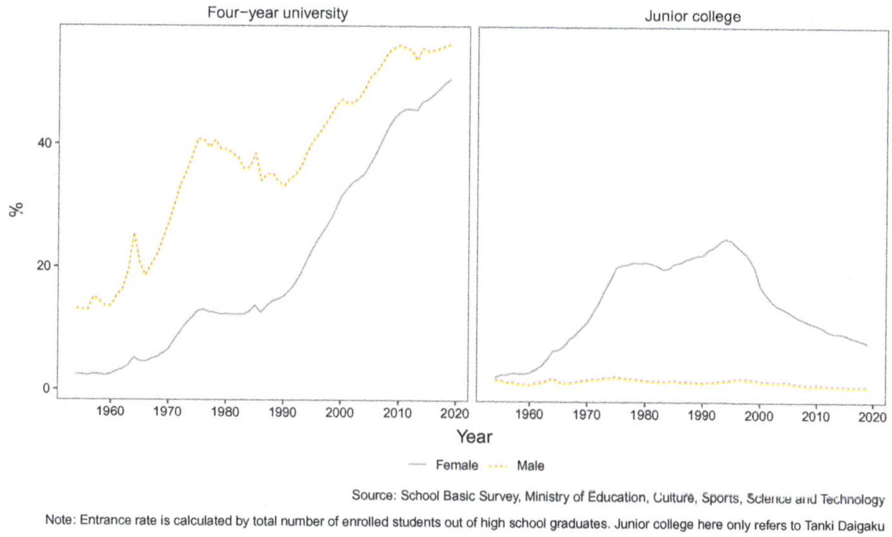

Fig. 4.1 Trends in enrollment rates for four-year universities and junior colleges

More specifically, if women's desire to marry hypergamously (with respect to educational attainment) remains strong (as Yamada (1996) suggests), relative improvements in women's status will shrink the pools of attractive mates for higher-status women and lower-status men. Indeed, the shortage of marriageable men preferred by highly educated women explains part of the decline in the proportion married in Japan (Raymo and Iwasawa 2005).

Assuming, for the sake of argument, that men and women's preference for female hypergamy has remained constant (an assumption that we question below), the convergence in men's and women's educational attainment is the driving force behind the marriage market mismatch just described. While gender differences in educational attainment have not disappeared, there has been a rapid convergence in recent years. Figure 4.1 shows trends in male and female entrance rates to four-year universities and junior colleges.[1] We can see that until recently, the likelihood of receiving post-secondary education was substantially greater for men than for women. During the first stage of college expansion from the late 1950s to the late 1970s, the proportion of students enrolled at the university level has increased for both men and women, but the increase was remarkable for men. The large majority of women who received higher education attended junior colleges, which are frequently characterized by their

[1] As this figure shows, the educational composition of the population has changed considerably, especially for women. The purpose of controlling compositional changes using log-linear models (by quantifying the contribution of marginal distributions) or harmonic mean models (by quantifying the availability ratio) is to adjust for the changing educational distribution and its potential influence homogamy patterns. By doing so, we can consider the role of other important social and economic factors, such as changing societal acceptance towards hypogamy, that is presumably reflected in the homogamy parameters in log-linear models or the force of attraction in harmonic mean models.

concentration of female students, constituting, for example, almost 90% of students enrolled in junior colleges during the 1980s (MEXT 2019). These colleges tended to offer traditionally female majors like education or home economics, reinforcing the ideal of the stay-at-home wife and mother (*ryōsai kenbo*) as an attractive marriage partner (Brinton 1988, 1992; Hijikata 1992). In fact, their emphasis on the "feminine arts" earned junior colleges the label of "bridal training schools" (*hanayome gakkō*) (e.g., Hamana 1990; Lebra 1984).

These gender differences in educational attainment perpetuated large gender differences in occupational status, with high schools and junior colleges preparing women for low-responsibility clerical positions in large firms, and thus the opportunity to participate in favorable marriage markets with low search costs. Until the collapse of the bubble economy and a series of deregulation policies, women generally remained in these OL (office lady) jobs only temporarily (*koshi kake*) before marriage, often to a man in the same company (*shokuba kekkon*) (Brinton 1992; McLendon 1983; Ogasawara 1998).

This distinctive pattern of women's education and employment no longer exists, as opportunities for women in higher education and the labor market have improved dramatically in recent decades. This change likely reflects both shifting attitudes and structural/policy changes. The Equal Employment Opportunity Act (*Danjo Koyō Kikai Kintōhō*, EEOA) passed in 1986 and revised in 1997, and the expansion of maternity and child care leave policies removed many of the formal obstacles that have perpetuated the temporary nature of women's employment. Indeed, several studies have suggested that changing social norms regarding women's employment following the passage of the EEOA encouraged women to have, and to pursue, career aspirations (Edwards and Pasquale 2003; Hashimoto 1992; Lam 1990).

These shifting norms and employment opportunities likely contributed to the increase in the number of women attending four-year universities. In 1990, for example, only 15% of female high school graduates went to four-year universities, a figure that increased rapidly, reaching 32% in 2019, with the gender gap in attendance at four-year universities shrinking to just 5 percentage points. Women's college majors have also changed in recent years. By the early 1990s, women attending four-year universities were increasingly entering traditionally male fields, such as the social sciences, engineering, and the physical sciences (Hijikata 1992; Makino 1995). It remains the case, however, that women outnumber men in humanities and other fields that are often thought of as feminine, while men are more likely to study science and engineering, resulting in the lack of female professionals in STEM (Science, Technology, Engineering, and Math) fields (Uchikoshi et al. 2020).

In the workplace too, career opportunities for women have improved markedly. Female labor force participation has changed considerably in recent years with the flattening of the M-shaped age curve (Brinton and Oh 2019). Indeed, female labor force participation in Japan surpassed that of the United States in 2016 (Shambaugh et al. 2017). We should keep in mind, however, that much of the increase in female labor force participation is explained by growth in relatively precarious, non-standard work (Nagase 2004; Tsutsui 2016) and the trend toward later and less marriage (Raymo and Fukuda 2016). Recent studies point to shifting educational differences in employment, with highly educated women more likely to remain in stable employment following marriage or childbearing (Kenjoh 2007; Lim and Raymo 2014; Raymo and Lim 2011; Raymo and Iwasawa 2016; Senda 2002) and accumulate human capital (Kawaguchi 2006), with implications for increasing wage/income inequality among female workers (Kambayashi et al. 2008).

4.4 Changing Context of Marriage Formation

Contextual influences on patterns of family formation have changed dramatically in Japan in recent decades. In the not-so-distant past, the family formation process was concentrated in a narrow band of ages (Brinton 1992) characterized by relatively strong facilitation of marriage, first by family members and then by workplace colleagues, often including one's immediate supervisor or boss. This third-party influence is perhaps most clearly represented by the involvement of parents in "arranged" marriages (*miai kekkon*), a method for meeting potential spouses that characterized more than one-quarter of marriages as late as the early 1980s. In a society characterized by an emphasis on family succession and lineage, arranged marriage had been used to select "appropriate" partners based on social origins (Shida et al. 2000; Yasuda 1971), but this form of marriage began a steady decline after WWII and only about 5% of marriages now fall into this category. As family facilitation of marriage waned, marriage among co-workers (*shokuba kekkon*), in which senior colleagues typically played an important role in facilitating matches among younger, single employees, increased in prevalence. But, as Iwasawa and Mita (2007) demonstrated, this type of marriage, and the role of the company as marriage market, has rapidly fallen out of favor. These changes in the nature of marriage markets, and third-party influence therein, can be seen as part of a larger de-standardization of the family life course in Japan. Not so long ago, Brinton (1992) described a family life course characterized by universal marriages, highly predictable timing of marriage, and the strict order of family formation, starting with courtship, marriage, and childbirth, but this is no longer an accurate picture of Japan's increasingly heterogenous family formation patterns.

4.4.1 From Arranged Marriage to Love Marriage

The shift from arranged, or "facilitated," marriage to love marriage was dramatic. Figure 4.2 presents the distribution of recently married couples (5-year marriage cohorts) by their marriage type (love marriage or arranged marriage). These data come from the National Fertility Survey, which is a regularly administered cross-sectional survey targeting single men and women and married women in Japan. In the oldest marriage cohort, about 7 out of 10 couples had an arranged marriage and "love marriage" described only 13% of couples. The prevalence of arranged and

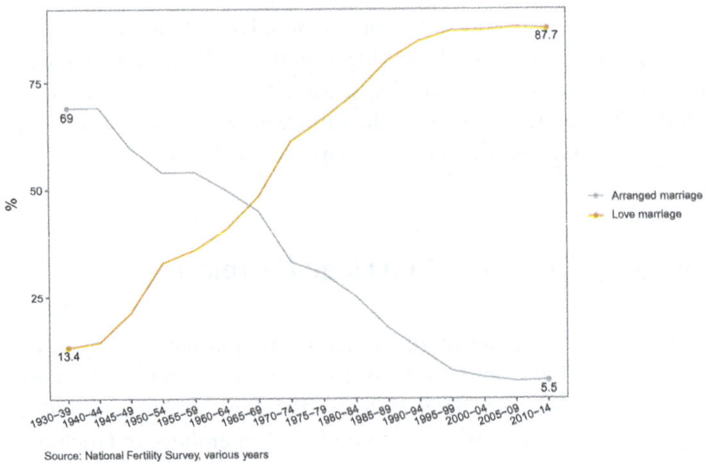

Fig. 4.2 Proportion of love marriages and arranged marriages

love marriages reversed around the 1960 marriage cohort, and from the 1990s and onwards, a majority (more than 85%) of married couples consider theirs to be a love marriage.

To understand trends in educational assortative mating, it is important to think about the implications of the declining role of third parties in facilitating the initial meeting of potential spouses. What does this shift mean for the marriage market and educational pairing? On the one hand, growth in the individual choice of partners may weaken social boundaries or normative expectations that may heighten the role of ascribed status in parents' (and others') identification of appropriate partners. In this case, we might expect to see a decline (or little change) in patterns of educational assortative mating. On the other hand, literature on assortative mating trends in the United States links the rise in educational homogamy (especially among the highly educated) to the role of schools as marriage markets and the increased importance of economic sorting (Mare 1991; Oppenheimer 1988; Schwartz and Mare 2005), suggesting that the decline in facilitated marriage may result in increased educational homogamy. Additionally, we suspect that the shift from arranged to love marriage promotes more educational hypogamy, a possibility suggested by recent research on China (Tian and Davis 2019). The logic of this argument is that a key role of third parties in introducing couples and facilitating marriages is to avoid non-normative marriages, which in the context of educational assortative mating in Japan have long included marriages in which the wife is more highly educated than the husband (i.e., educational hypogamy).

4.4.2 *Changing Composition of Love Marriage*

While we believe that the distinction between arranged marriage and love marriage is critical for understanding changes in assortative mating, we also recognize that the simple dichotomy between love marriage and arranged marriage likely masks a great deal of heterogeneity, especially in the former category. In particular, couples in love marriages report meeting their spouse in a wide array of settings, information that may provide valuable insights into patterns of educational selection in the marriage market. Importantly, previous studies have found that meetings in the workplace, which might be seen as "lightly facilitated" marriages (Iwasawa and Mita 2007), have been replaced by meeting via friends or meeting at school, both of which are presumably less influenced by parental, or other third-party, intervention (Iwasawa 2013). Assuming that some meetings through friends are qualitatively similar to meeting at school (e.g., classmates introduce a friend), the growth in marriages emerging from these types of meetings should increase the prevalence of educational homogamy. According to Mare (1991), if we hypothetically assume marriage timing to be constant, increasing educational attainment will lead to more educational homogamy because it reduces the time gap between school completion and marriage (i.e., it increases the likelihood of meeting one's spouse while in school). For all of these reasons, we speculate that the changing nature of social interaction, dating, and marriage (i.e., the dating and mating "market") should result in an increase in educational homogamy. We also reiterate that, here and throughout this discussion, we are referring to the likelihood of educational homogamy (or other types of educational pairing) net of changes in the distribution of educational attainment. That is, we are talking about behavior within the marriage market rather than the composition of the marriage market. Both are important, but it is critical to distinguish the ways in which each shapes patterns of spouse pairing.

4.4.3 *Changing Timing of Marriage and the Length of Relationship*

Importantly, Mare (1991) also argued that an increase in age at marriage may reduce the odds of educational homogamy. The longer the time gap between school completion and marriage, the more likely it is that unmarried men and women will meet their spouse outside of the school context, pairings which are more likely to be educationally heterogamous (Kalmijn 1998). It is well established that the age at first marriage in Japan has increased substantially. According to vital statistics data, the mean age at first marriage in 1975 was 27.0 for men and 24.7 for women, and increased to 31.2 for men and 29.6 for women by 2019. Although data on the average age at school completion are not available, the university entrance rate increased by 15% for men and 38% for women during the same period. Under some simple assumptions about the average age of completing different levels of education, this growth in university attendance would increase the age of school completion by 0.6 years for men and 1.3 years for women. This change is substantially smaller than the increase in age at first marriage. Although we understand that shrinking cohort size should also be considered, this quick calculation makes clear that the increase in age at first marriage has been much faster than the increase in age at school completion, suggesting a likely decline in educational homogamy in recent decades.

Another important factor influencing patterns of marriage timing and spouse pairing is the growing length of dating relationships in recent years. According to data from the NFS, the average length of relationships prior to marriage (dating and engagement, possibly including a period of cohabitation) among first married couples increased from 2.5 years in 1975 to 4.3 years in 2015. Even if we limit the sample to love marriages (because the length of the pre-marital relationship for arranged marriages tends to be shorter), the average length of the dating relationship prior to marriage increased by 1.4 years over this 40-year period. We speculate that this trend may mitigate the posited negative impact of the growing gap between school completion and marriage on the prevalence of educational homogamy.

4.4.4 Changing Spouse Selection Criteria

Japan's decades-long economic stagnation and increasing labor market uncertainty characterized by growth in non-standard employment and decline in long-term employment appears to be reflected in changing spouse selection criteria. More specifically, there is evidence of increasing gender symmetry in emphasis on the earnings potential of potential spouses. Women have long prioritized potential husbands' earnings potential and it now appears that single men, who increasingly hope and expect their wives to work, are placing more emphasis on women's earnings potential (Brinton et al. 2021; IPSS 2017). One key factor in this attitudinal shift is the relative stability of returns to educational attainment in the Japanese labor market (Kawaguchi and Mori 2016; Yamada and Kambayashi 2015). In contrast to many other countries where growing educational differences in wages contribute to growing economic inequality (Acemoglu and Autor 2011), the stable wage gap and an increasing educational gap in work hours in Japan (Toyonaga 2021) suggests that growing incentives to marry highly educated men seen in the United States (Schwartz and Mare 2005) may be less relevant for understanding trends in educational assortative mating in Japan. At the same time, general wage stagnation and employment uncertainty have led many single men and women to increasingly believe that the male breadwinner-female homemaker model of marriage is no longer feasible (IPSS 2017). According to the most recent NFS data (collected in 2015), the proportion of single women expressing a desire to be a full-time housewife has declined consistently over time as the proportion hoping to balance work and family has grown. Single men also increasingly indicate a desire for their wife to work and place more importance on women's earnings potential as a spouse selection criterion. These attitudinal shifts suggest a possible decline in preferences for female hypergamy and recent studies show that women with higher economic potential (as measured by earnings and education) have a higher likelihood of marrying (Fukuda 2013; Fukuda et al. 2020). We believe this is an important trend that may contribute to a subsequent increase in educational hypogamy in Japan (Fukuda et al. 2021).

References

Acemoglu, Daron, and David Autor. 2011. Skills, tasks and technologies: Implications for employment and earnings. *Handbook of Labor Economics* 1043–1171.

Blossfeld, Hans-Peter, and Andreas Timm. 2003. *Who marries whom?* Springer.

Brinton, Mary C. 1988. The social-institutional bases of gender stratification: Japan as an illustrative case. *American Journal of Sociology* 94 (2): 300–334. https://doi.org/10.1086/228993.

Brinton, Mary C. 1991. Sex differences in on-the-job training and job rotation in Japanese Firms. *Research in Social Stratification and Mobility* 10: 3–25.

Brinton, Mary C. 1992. Christmas cakes and wedding cakes: The social organization of Japanese women's life course. In *Japanese social organization*, ed. T. S. Lebra, 79–107. University of Hawaii Press.

Brinton, Mary C. 1993. *Women and the economic miracle: Gender and work in postwar Japan*. University of California Press.

Brinton, Mary C. 2001. *Women's working lives in East Asia*. Stanford University Press.

Brinton, Mary C. 2007. Gendered offices: A comparative-historical examination of clerical work in Japan and the United States. In *Political economy of Japan's low fertility*, ed. F. M. Rosenbluth, 87–111. Stanford University Press.

Brinton, Mary C., and Eunsil Oh. 2019. Babies, work, or both? Highly educated women's employment and fertility in East Asia. *American Journal of Sociology* 125 (1): 105–140. https://doi.org/10.1086/704369.

Brinton, Mary C., and Sunhwa Lee. 2001. Women's education and the labor market in Japan and South Korea. In *Women's working lives in East Asia*, ed. M. C. Brinton, 125–50. Stanford University Press.

Brinton, Mary C., Eunmi Mun, and Ekaterina Hertog. 2021. Singlehood in contemporary Japan: Rating, dating, and waiting for a good match. *Demographic Research* 44: 239–276. https://doi.org/10.4054/DemRes.2021.44.10.

Bumpass, Larry L., Ronald R. Rindfuss, Minja Choe, and Noriko O. Tsuya. 2009. the institutional context of low fertility: The case of Japan. *Asian Population Studies* 5 (3): 215–235. https://doi.org/10.1080/17441730903351479.

Chase, Ivan D. 1975. A comparison of men's and women's intergenerational mobility in the United States. *American Sociological Review* 40 (4): 483–505.

Cheng, Yuan, and Jianzhong Dai. 1995. Intergenerational mobility in modern China. *European Sociological Review* 11 (1): 17–35. https://doi.org/10.1093/oxfordjournals.esr.a036347.

Clark, Robert L., and Naohiro Ogawa. 1992. Employment tenure and earnings profiles in Japan and the United States: Comment. *American Econmic Review* 82 (1): 336–345.

Edwards, Linda N., and Margaret K. Pasquale. 2003. Women's higher education in Japan: Family background, economic factors, and the equal employment opportunity law. *Journal of the Japanese and International Economies* 17 (1): 1–32.

Erikson, Robert, and John H. Goldthorpe. 1992. *The constant flux: A study of class mobility in industrial societies*. Oxford University Press.

Estévez-Abe, Margarita. 2008. *Welfare and capitalism in postwar Japan*. Cambridge University Press.

Fujihara, Sho, and Fumiya Uchikoshi. 2019. Declining association with persistent gender asymmetric structure: Patterns and trends in educational assortative marriage in Japan, 1950–1979.

Research in Social Stratification and Mobility 60: 66–77. https://doi.org/10.1016/j.rssm.2018.12.001.

Fujimi, Junko. 2009. Fufu No Katachi, Kekkon No Katachi (Forms of Couples and Marriage). In *Gendai Nihonjin no Kazoku (Families in Contemporary Japan)*, ed. J. Fujimi and M. Nishino, 55–71. Yuhikaku. (in Japanese).

Fukuda, Setsuya. 2013. The changing role of women's earnings in marriage formation in Japan. *The ANNALS of the American Academy of Political and Social Science* 646 (1): 107–128. https://doi.org/10.1177/0002716212464472.

Fukuda, Setsuya, James M. Raymo, and Shohei Yoda. 2020. Revisiting the educational gradient in marriage in Japan. *Journal of Marriage and Family* 82 (4): 1378–1396. https://doi.org/10.1111/jomf.12648.

Fukuda, Setsuya, Shohei Yoda, and Ryohei Mogi. 2021. Educational assortative mating in Japan: Evidence from the 1980–2010 census. *The Journal of Population Studies (Jinkogaku Kenkyu)*. https://doi.org/10.24454/jps.2101001.

Glenn, Norval D., Adreain Ross, and Judy Tully. 1974. Patterns of intergenerational mobility of females through marriage. *American Sociological Review* 39 (5): 683. https://doi.org/10.2307/2094314.

Hamana, Atsushi. 1990. Josei Ni Okeru Gakko Riyoso No Bunseki (An Analysis of Female Educational Participation). In *Gendai Nihon no Kaiso Kozo 3: Kyoiku to Syakai Ido (Class Structure in Modern Japan 3: Eucation and Social Mobility)*, ed. J. Kikuchi, 85–106. University of Tokyo Press. (in Japanese).

Hara, Toshihiko. 2008. Increasing childlessness in Germany and Japan: Toward a childless society? *International Journal of Japanese Sociology* 17 (1): 42–62. https://doi.org/10.1111/j.1475-6781.2008.00110.x.

Hashimoto, Noriko. 1992. *Danjo Kyogakusei No Shiteki Kenkyu (A Historical Study of Mixed-Sex Schools)*. Otsuki Shobo. (in Japanese).

Heath, Anthony. 1981. *Social mobility*. Harper Collins.

Hijikata, Utsuko. 1992. Josei to Kyoiku (Women and Education). In *Josei no Raifusaikuru (Women's Life Cycle)*, ed. M. Ozawa, S. Kimura, and H. Ibe, 13–35. University of Tokyo Press. (in Japanese).

Hirao, Keiko. 2001. Mothers as the best teachers: Japanese motherhood and early childhood education. In *Women's working lives in East Asia*, ed. M. C. Brinton, 180–203. Stanford University Press.

Holloway, Susan D. 2010. *Women and family in contemporary Japan*. Cambridge University Press.

Hashimoto, Masanori, and John Raisian. 1985. Employment tenure and earnings profiles in Japan and the United States. *American Economic Review* 75 (4): 721–735.

Ishida, Hiroshi, Su Kuo-hsien, and Seymour Spilerman. 2002. Models of career advancement in organizations. *European Sociological Review* 18 (2): 179–198. https://doi.org/10.1093/esr/18.2.179.

Iwasawa, Miho. 1999. The state of women's life courses in contemporary Japan: Focusing on never-married women's prospects. *Journal of Population Problems (Jinko Mondai Kenkyu)* 55 (4): 16–37 (in Japanese).

Iwasawa, Miho. 2013. An analysis of marriage decline and changes in marriage structures in Japan since the 1970s using multiple decrement life tables on first marriage. *Journal of Population Problems (Jinko Mondai Kenkyu)* 69 (2): 1–34 (in Japanese).

Iwasawa, Miho, and Fusami Mita. 2007. Boom and bust in marriages between coworkers and the marriage decline in Japan. *Japanese Economy* 34 (4): 3–24. https://doi.org/10.2753/JES1097-203X340401.

Kalleberg, Arne L., and James R. Lincoln. 1988. The structure of earnings inequality in the United States and Japan. *American Journal of Sociology* 94: S121–S153. https://doi.org/10.1086/228944.

Kalmijn, Matthijs. 1998. Intermarriage and homogamy: Causes, patterns, trends. *Annual Review of Sociology* 24: 395–421.

Kambayashi, Ryo, Daiji Kawaguchi, and Izumi Yokoyama. 2008. Wage distribution in Japan, 1989–2003. *Canadian Journal of Economics/revue Canadienne D'économique* 41 (4): 1329–1350. https://doi.org/10.1111/j.1540-5982.2008.00506.x.

Kawaguchi, Daiji. 2006. The incidence and effect of job training among Japanese women. *Industrial Relations* 45 (3): 469–477. https://doi.org/10.1111/j.1468-232X.2006.00433.x.

Kawaguchi, Daiji, and Yuko Ueno. 2013. Declining long-term employment in Japan. *Journal of the Japanese and International Economies* 28: 19–36. https://doi.org/10.1016/j.jjie.2013.01.005.

Kawaguchi, Daiji, and Yuko Mori. 2016. Why has wage inequality evolved so differently between Japan and the US? The role of the supply of college-educated workers. *Economics of Education Review* 52: 29–50. https://doi.org/10.1016/j.econedurev.2016.01.002.

Kenjoh, Eiko. 2007. Employment options: Japan in comparative perspective. In *The political economy of Japan's low fertility*, ed. F. M. Rosenbluth, 112–30. Stanford University Press.

Lam, Alice. 1990. Equal employment opportunities for Japanese women: Changing company practice. London School of Economics and Political Science.

Lebra, Takie Sugiyama. 1984. *Japanese women: Constraint and fulfillment.* University of Hawai'i Press.

Lim, So-Jung, and James M. Raymo. 2014. Nonstandard work and educational differentials in married women's employment in Japan: Patterns of continuity and change. *International Journal of Sociology* 44 (3): 84–107. https://doi.org/10.2753/IJS0020-7659440305.

Makino, Fumio. 1995. Joshi No Kogakurekika to Shokugyo Sentaku (Womens Increasing Educational Attainment an Occupational Choice). In *Josei no Raifusaikuru to Shugyo Kodo (Womens Life Cycle and Employment Behavior)*, ed. H. Ohbuchi, 37–59. Okurasho Insatsukyoku (in Japanese).

Mare, Robert D. 1991. Five decades of educational assortative mating. *American Sociological Review* 56 (1): 15–32.

Mclendon, James. 1983. The office: Way station or blind alley? In *Work and lifecourse in Japan*, ed. by D. W. Plath, 156–82. State University of New York Press.

Ministry of Education 2019. Culture Sports, Science and Technology. *Statistical Abstract*.

Mincer, Jacob, and Yoshio Higuchi. 1988. Wage structures and labor turnover in the United States and Japan. *Journal of the Japanese and International Economies* 2 (2): 97–133. https://doi.org/10.1016/0889-1583(88)90017-2.

Nagase, Nobuko. 2004. Standard and nonstandard work arrangements, pay differences, and choice of work by Japanese Mothers. In *Nonstandard work in developed economies: Causes and consequences*, ed. S. Houseman and M. Osawa, 267–306. W.E. Upjohn Institute.

National Institute of Population and Social Security Research. 2017. *Marriage and Childbirth in Japan Today: The Fifteenth Japanese National Fertility Survey, 2015*. National Institute of Population and Social Security Research.

Nemoto, Kumiko. 2016. *Too few women at the top: The persistence of inequality in Japan.* ILR Press, an imprint of Cornell University Press.

Ogasawara, Yuko. 1998. *Office ladies and salaried men: Power, gender and work in Japanese companies*. University of California Press.

Ogawa, Naohiro, and Robert L. Clark. 1995. Earnings patterns of Japanese women: 1976–1988. *Economic Development and Cultural Change* 43 (2): 293–313. https://doi.org/10.1086/452151.

Oppenheimer, Valerie Kincade. 1988. A theory of marriage timing. *American Journal of Sociology* 94 (3): 563–591.

Osawa, Machiko. 1993. *Keizai Henka to Joshi Rodo (Economic Change and Women's Work)*. Nihon Keizai Shimbunsha.

Osawa, Mari. 1993. *Kigyo Chushin Shakai Wo Koete (Beyond the Firm-Oriented Society)*. Jijitsushinsha (in Japanese).

Portocarero, Lucienne. 1985. Social mobility in France and Sweden: Women, marriage and work. *Acta Sociologica* 28 (3): 151–170. https://doi.org/10.1177/000169938502800301.

Raymo, James M., Hyunjoon Park, Yu Xie, and Wei-jun Jean Yeung. 2015. Marriage and family in East Asia: Continuity and change. *Annual Review of Sociology* 41:471–92. https://doi.org/10.1146/annurev-soc-073014-112428.

Raymo, James M., and Miho Iwasawa. 2005. Marriage market mismatches in Japan: An alternative view of the relationship between women's education and marriage. *American Sociological Review* 70 (5): 801–822. https://doi.org/10.1177/000312240507000504.

Raymo, James M., and Miho Iwasawa. 2016. *Diverging destinies: The Japanese case*. Springer.

Raymo, James M., and Setsuya Fukuda. 2016. Trends in women's labor force participation: The role of changing marital behavior. *The Japanese Journal of Labour Studies* 674: 26–38.

Raymo, James M., and So-Jung Lim. 2011. A new look at married women's labor force transitions in Japan. *Social Science Research* 40 (2): 460–472. https://doi.org/10.1016/j.ssresearch.2010.10.005.

Rossi, Alice. 1965. Women in science: Why so few? *Nature* 3674: 1196–1202.

Schwartz, Christine R. 2013. Trends and variation in assortative mating: Causes and consequences. *Annual Review of Sociology* 39 (1): 451–470. https://doi.org/10.1146/annurev-soc-071312-145544.

Schwartz, Christine R., and Robert D. Mare. 2005. Trends in educational assortative marriage from 1940 to 2003. *Demography* 42 (4): 621–646.

Senda, Yukiko. 2002. Influence of childcare resources on the employment continuity of married women: Focusing on occupation and birth cohort. *Journal of Population Problems (Jinko Mondai Kenkyu)* 58 (2): 2–21 (in Japanese).

Shambaugh, Jay, Ryan Nunn, and Becca Portman. 2017. Lessons from the rise of women's labor force participation in Japan. *The Brookings Institution.*

Shida, Kiyoshi, HIdeki Watanabe, and Kazuo Seiyama. 2000. Kekkon Shijo No Henyo (Changes in Marriage Market). In *Nihon no Kaiso System: Gender, Shijo, Kazoku (Stratification System in Japan: Gender, Market, and Family)*, ed. K. Seiyama, 159–76. University of Tokyo Press. (in Japanese).

Shirahase, Sawako. 2001. Women and class structure in contemporary Japan. *The British Journal of Sociology* 52 (3): 391–408. https://doi.org/10.1080/00071310120071115.

Takeuchi, Mamiko. 2004. Josei Shugyo No Paneru Bunseki: Haigusha Shotoku Koka No Saikensho (Panel Analysis of Women's Decision to Work: Re-Examination of the Effects of the Husband's Income). *The Japanese Journal of Labour Studies* 46(6):76–88 (in Japanese).

Tian, Felicia F., and Deborah S. Davis. 2019. Reinstating the family: Intergenerational influence on assortative mating in China. *Chinese Sociological Review* 51 (4): 337–364. https://doi.org/10.1080/21620555.2019.1632701.

Toyonaga, Kohei. 2021. Kosotsu to Daisotsu No Gakureki Kan Chingin Kakusa Wa Kakudai Shitanoka? (Has Wage Gap Between High School Graduates and University Graduates Increased?). *Kikan Kojin Kinyu (quarterly Personal Finance)* 15 (4): 78–87 (in Japanese).

Tsuya, Noriko O, and Minja Kim, Choe. 2004. Investments in children's education, desired fertility, and women's employment. In *Marriage, work, and family in comparative perspective*, ed. N. O. Tsuya and L. L. Bumpass, 76–94. University of Hawai'i Press.

Tsutsui, Junya. 2016. *Kekkon to Kazoku No Korekawa (The Future of Marriage and Family).* Kobunsha. (in Japanese).

Tsutsui, Junya. 2020. *Work and family in Japanese society*. Springer Singapore.

Tyree, Andrea, and Judith Treas. 1974. The occupational and marital mobility of women. *American Sociological Review* 39 (3): 293–302.

Uchikoshi, Fumiya, Ryota Mugiyama, and Megumi Oguro. 2020. Still separate in STEM? Trends in sex segregation by field of study in Japan, 1975–2019. *Hitotsubashi University Institute of Economic Research Working Paper Series A* 710.

van Bavel, Jan, Christine R. Schwartz, and Albert Esteve. 2018. The reversal of the gender gap in education and its consequences for family life. *Annual Review of Sociology* 44:341–60. https://doi.org/10.1146/annurev-soc-073117-041215.

Yamada, Ken, and Daiji Kawaguchi. 2015. The changing and unchanged nature of inequality and seniority in Japan. *The Journal of Economic Inequality* 13 (1): 129–153. https://doi.org/10.1007/s10888-014-9295-6.

Yamaguchi, Kazuo. 2019. *Gender inequalities in the Japanese workplace and employment: Theories and empirical evidence*. Vol. 22. Springer Singapore.

Yamada, Masahiro. 1996. *Kekkon No Shakaigaku (The Sociology of Marriage)*. Maruzen. (in Japanese).

Yasuda, Saburo. 1971. *Shakai Ido No Kenkyu (A Study of Social Mobility)*. University of Tokyo Press. (in Japanese).

Yu, Wei-hsin. 2006. National contexts and dynamics of married women's employment reentry: The cases of Japan and Taiwan. *The Sociological Quarterly* 47 (2): 215–243. https://doi.org/10.1111/j.1533-8525.2006.00044.x.

Yu, Wei-hsin. 2009. *Gendered trajectories*. Stanford University Press.

Zhang, Shiying. 2012. Kikon Josei No Rodo Kyokyu to Otto No Shotoku (Labor Supply of Married Women and Husband's Earnings). *Kikan Shakai Hosho Kenkyu (Quarterly Journal of Social Security)* 47 (4): 401–412 (in Japanese).

Zhou, Yanfei. 2015. Career interruption of Japanese women: Why is it so hard to balance work and childcare? *Japan Labor Review* 12 (2): 106–123.

Chapter 5
Empirical Analysis

5.1 Introduction

In this chapter, we consider two broad empirical questions—one about the process of assortative mating and the other about its consequences. The first question covers two interrelated, but independent, sub-questions. First, we examine the union formation and marriage process. Specifically, we ask whether patterns of educational assortative mating/pairing differ between dating, cohabiting, and married couples. Furthermore, we evaluate the winnowing hypothesis by focusing on the length of dating relationships, hypothesizing that the longer the dating relationship, the more educationally homogamous couples will be. Second, we examine patterns and trends of assortative mating more broadly by expanding the definition of assortativity. Specifically, we examine assortative mating by both educational attainment and age and also consider the notion of status exchange. The second set of questions investigates the consequences of educational assortative mating, focusing on women's labor force participation patterns after marriage as well as trends in income inequality in Japan. We then examine how educational pairing is related to fertility, divorce, and spousal division of labor. Lastly, we examine questions about the potential relevance of educational assortative mating for intergenerational mobility by focusing on differences in expenditures on children's education by parents' educational pairing.

5.2 Data and Analysis Plan

In this chapter, we use two nationally representative two datasets—SSM (Social Stratification and Mobility Study) and JLPS (Japanese Life-course Panel Study). We also use vital statistics and census data to describe patterns of assortative mating with respect to age. We begin by summarizing the basic characteristics of each data source. Ideally, we would like to use information from large-scale administrative data sources, but access to individual-level data from the census and vital statistics

F. Uchikoshi and J. M. Raymo, *Educational Assortative Mating in Japan*,
Population Studies of Japan, https://doi.org/10.1007/978-981-16-3713-1_5

is limited and information on educational attainment is collected only at 10-year intervals in the census and not at all in the vital statistics. We can use the vital statistics and census data for analyses of assortative mating with respect to age but need to rely on survey data for all other analyses. SSM is one of the oldest social surveys in Japan, starting in 1955 and conducted every 10 years through 2015. This survey's primary focus is on occupational mobility within and across generations, but it also includes information on respondents' education and spouses' socioeconomic characteristics, including educational attainment. The SSM began to include female respondents in 1985, so we use data from the four most recent surveys: SSM 1985, 1995, 2005, and 2015.

In addition to these cross-sectional data, we also use a longitudinal dataset that tracks the same individuals over time. JLPS is a nationally representative longitudinal survey of young and middle-aged men and women (aged 20–40 at the time of the initial survey), conducted annually by the Institute of Social Science at the University of Tokyo. Data are currently available from wave 1 (2007) to wave 9 (2015). Ishida (2013) provides detailed information on the data structure and sampling procedures. One strength of JLPS data is its collection of detailed relationship histories including the length of dating relationships before marriage and where/how couples met. These data allow us to examine the union formation and marriage process in a level of detail that facilitates the testing of several important hypotheses discussed in previous research on educational assortative mating.

5.3 Formation of Educationally Assortative Mating

5.3.1 The Marriage Process

Previous studies have reached mixed conclusions regarding the winnowing hypothesis. We test this hypothesis using data drawn from the Japanese Life-course Panel Study (JLPS). We build upon and extend in important ways recent studies by Ishida and Motegi (2014) and Motegi and Ishida (2020) that examined this question using log-linear models to describe patterns of educational pairing by relationship type (dating and marriage). Our analyses examine cohabiting couples, a union type that is growing in prevalence and has been described as a precursor to marriage rather than an alternative to marriage (Raymo et al. 2009). As a step in the marriage process, cohabitation should theoretically fall somewhere between dating and marriage in terms of the educational resemblance of partners. As such, comparison of educational pairing patterns across these relationship types can provide important insights regarding the winnowing hypothesis. We also examine educational homogamy as an outcome whereas Motegi and Ishida's log-linear approach described the likelihood of educational homogamy relative to other types of spouse pairing patterns. One drawback of using log-linear models to examine contingency tables is the difficulty of incorporating theoretically relevant covariates. In testing the winnowing hypothesis, one such variable is relationship duration, a continuous measure that would be particularly difficult to incorporate in a log-linear framework. We therefore estimate logistic regression models in which the outcome of interest is whether the respondent's romantic union is educationally homogamous or not. Independent variables include relationship duration as well as respondent's educational attainment and relationship type (dating, cohabitation, marriage). Summary statistics for all variables used in the analyses are shown in Table 5.1.

Our sample includes (1) individuals who are in a dating relationship at the time of the survey, (2) those who are in a cohabiting relationship at the time of the survey, and (3) those who married during the past year (newlyweds). We limit the third category to those who married for the first time in light of evidence that spouse selection mechanisms and pairing patterns for second and higher-order marriages are different from those of first marriages (Qian and Lichter 2018; Shafer 2013). For the first and second groups, we include multiple observations of the same individuals to increase the sample size. We assume that those observed more than once in a dating/cohabiting relationship are with the same partner, but the data do not allow us to evaluate this assumption. To account for repeated observations of the same individuals, we used clustered standard errors.

Table 5.2 presents the results of these regression models. The first three models included all respondents regardless of relationship type. Compared with middle-educated (junior college graduates), we see that both low- and high-educated individuals are more likely to be in educationally homogamous relationships. These results are similar to findings from previous studies showing higher odds of homogamy for the least- and most-educated groups relative to those in the middle of the educational distribution (Fujihara and Uchikoshi 2019; Fukuda et al. 2021). We also see that, consistent with one prediction of the winnowing hypothesis, relationship duration is positively associated with the odds of homogamy, suggesting that educationally

Table 5.1 Descriptive statistics

	Mean	SD	Min	Max
Educational homogamy	0.53	0.50	0.00	1.00
Female	0.61	0.49	0.00	1.00
High school or less	0.25	0.43	0.00	1.00
Junior college	0.30	0.46	0.00	1.00
BA +	0.45	0.50	0.00	1.00
Age	33.11	6.04	24.00	49.00
Dating without cohabiting	0.60	0.49	0.00	1.00
Cohabitation (incl. engaged)	0.06	0.24	0.00	1.00
Newly wed	0.34	0.47	0.00	1.00
Duration of relationship (months)	55.47	59.94	1.00	340.00
Observations	2026			

Table 5.2 Regression results for models of educational homogamy

	Model 1		Model 2		Model 3		Model 4		Model 5		Model 6	
Sample	Total		Total		Total		Dating		Cohabiting		Newly wed	
Female	0.117	(0.138)	0.112	(0.138)	0.108	(0.138)	0.045	(0.176)	− 0.332	(0.472)	0.345*	(0.171)
High school or less	1.031***	(0.171)	1.258***	(0.211)	1.024***	(0.171)	1.213***	(0.225)	0.517	(0.603)	0.802***	(0.219)
BA +	1.333***	(0.158)	1.335***	(0.158)	1.194***	(0.185)	1.278***	(0.201)	0.963	(0.599)	1.506***	(0.198)
Age	−0.007	(0.012)	−0.008	(0.012)	−0.008	(0.012)	−0.007	(0.015)	−0.011	(0.045)	−0.005	(0.016)
Cohabitation (incl. engaged)	−0.218	(0.248)	−0.107	(0.305)	−0.185	(0.325)						
Newly wed	−0.126	(0.115)	0.021	(0.125)	−0.328*	(0.159)						
Duration of relationship (months)	0.003*	(0.001)	0.003*	(0.001)	0.003*	(0.001)	0.004*	(0.002)	0.016**	(0.005)	0.001	(0.001)
High school or less # Cohabitation (incl. engaged)			−0.378	(0.518)								
High school or less # Newly wed			−0.600*	(0.243)								
BA + # Cohabitation (incl. engaged)					−0.123	(0.487)						
BA + # Newly wed					0.425*	(0.204)						
Constant	−0.666	(0.429)	−0.696	(0.430)	−0.572	(0.432)	−0.709	(0.544)	−0.869	(1.703)	−0.901	(0.552)
Observations	2026		2026		2026		1212		128		686	
Log likelihood	−1321.679		−1318.178		−1319.251		−786.161		−79.961		−443.091	
Standard errors in parentheses												

+ $p < 0.10$ * $p < 0.05$ ** $p < 0.01$ *** $p < 0.001$

homogamous couples are less likely to break up. Interestingly, the odds of homogamy are not statistically different across relationship types.

One limitation of this baseline model is the assumption that coefficients for relationship type are the same across different educational levels. To examine potential heterogeneity by union type, we estimated interactions between relationship type and educational attainment. Interestingly, the relative odds of homogamy by education level are different among newlyweds and those who are dating. Relative to those who are dating, newly married individuals in the lowest educational group are less likely to pair homogamously, whereas those in the highest educational category are more likely to mate homogamously. These contrasting results can be understood in the context of shifting economic foundations of marriage (Sweeney 2002), with a valuation of women's earnings potential incentivizing homogamy among the highly educated to a greater degree than among those with less education (i.e., lower earnings potential). We also note that we cannot reject the hypothesis that cohabiting couples are statistically different from dating couples in terms of the likelihood of educational homogamy. This is somewhat surprising in light of the aforementioned evidence that cohabiting unions in Japan are often a prelude to marriage (Raymo et al. 2009).

To further examine heterogeneity in educational pairing by relationship type, we estimated the same models separately for dating (Model 4), cohabiting (Model 5), and married individuals (Model 6). Results show that low- and high-educated individuals are both more likely to pair with similarly educated partners in dating relationships and marriage. The fact that we do not see a similar pattern for cohabiting couples may simply reflect the smaller sample size of this group. Indeed, coefficients for the low- and high-educated groups are both positive, although not significantly different from zero, in the model for cohabitors. Looking at educational differences within pairing type, we see that the positive coefficients for educational homogamy are similar between the low- and high-educated daters, while results for newlyweds indicate that highly educated individuals are more likely to mate homogamously than are those in a dating relationship.

5.3.2 *Who Marries Whom?: Patterns and Trends*

5.3.2.1 Assortative Mating by Education

There is abundant evidence to suggest that educational homogamy in Japan has declined both in absolute and relative terms (Fujihara and Uchikoshi 2019; Fukuda et al. 2021; Miwa 2007; Raymo and Xie 2000). Our aim in this section is to use SSM data to revisit some of the basic questions examined in previous research on educational assortative mating in Japan. One innovation is our focus on marriage cohorts rather than birth cohorts—a preferable definition of cohort for examining experiences in the marriage market. We also extend previous research on educational assortative mating by paying particular attention to trends in educational hypogamy, a key focus of recent research on trends around the world (Esteve et al. 2012, 2016) and the role played by the reversal in the gender gap in higher education (van Bavel 2012; van Bavel et al. 2018).

Our analytical sample is comprised of those who were married (for 10 years or less) at the time of each survey and married between 1975 and 2015. Based on the information about age at marriage, we constructed 10-year marriage cohorts. Although we use a three-category measure of educational attainment in other chapters (reflecting the small number of junior high school graduates in recent years), here we separate junior high school and high school graduates given the sizable number in the former group among earlier cohorts and evidence that the strength of educational homogamy differs between the two groups (Fujihara and Uchikoshi 2019; Fukuda et al. 2021). To make our results comparable to previous studies, we use log-linear and log-multiplicative models (see Chap. 3 for details of each model).

Figure 5.1 presents the distribution of educational pairings across marriage cohorts. This figure clearly shows that educational homogamy has declined while educational hypogamy increased. These trends in the prevalence of specific types of educational pairing have implications for sociodemographic outcomes such as income inequality (Breen and Andersen 2012; Breen and Salazar 2010; Greenwood et al. 2014; Hu and Qian 2015; Monaghan 2015; Schwartz 2010; Torche 2010; Western et al. 2008; Zagel and Breen 2019), but because they reflect both compositional shifts in educational attainment and spouse pairing behavior, it is important that we also estimate the odds, or relative likelihood, of different pairings net of marginal distributions.

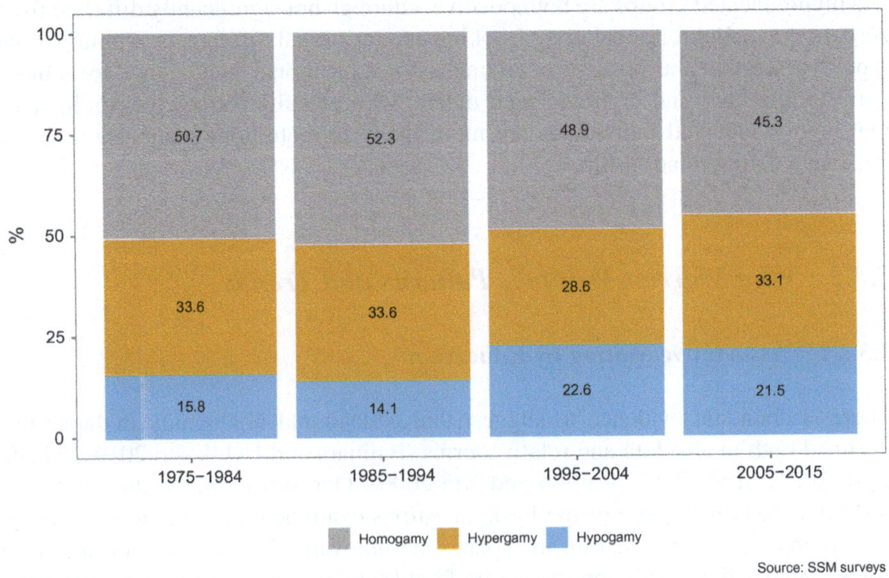

Fig. 5.1 Percent of couples by educational pairing type

Table 5.3 presents the fit statistics for log-linear and log-multiplicative models of educational pairing. Details about each model can be found in the Appendix. Although our primary aim is not to identify the best fitting model, these fit statistics suggest that Model 5, which assumes that educational hypergamy has changed over time, fits the data best, followed by Model 7, which assumes uniform cohort change in the degree of assortative mating (i.e., the UNIDIFF model).

Before describing cohort trends, we examine the strength of homogamy at each education level, as estimated by Model 2. Although this model does not fit the data as well as some other models, it is useful for comparing our estimates of differences in pairing propensities by educational attainment with those from previous studies. Figure 5.2 presents the coefficients and their 95% confidence intervals. As found in previous studies (Fujihara and Uchikoshi 2019; Fukuda et al. 2021), the strength of

Table 5.3 Model fit statistics

	df	G^2	Index of Dissimilarity	BIC
Model 1	36	809.8	0.2	521.7
Model 2	32	204	0.07	−52
Model 3	15	34.7	0.02	−85.3
Model 4	24	51.2	0.03	−140.8
Model 5	24	47.8	0.03	−144.2
Model 6	24	48.3	0.03	−143.7
Model 7	24	48.1	0.03	−143.9

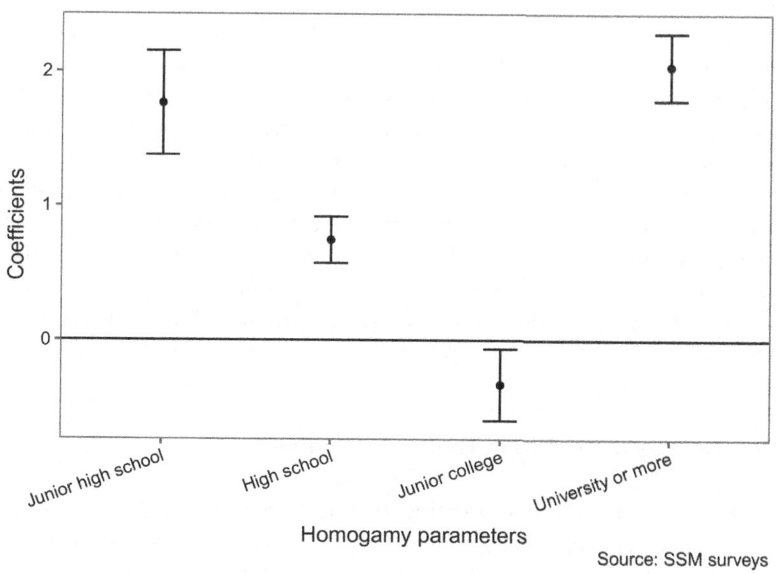

Fig. 5.2 Visualization of homogamy coefficients (coarse categories for university graduates)

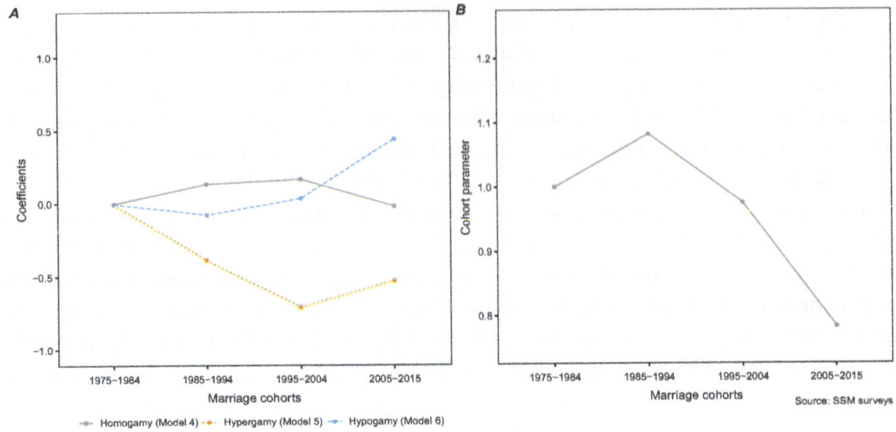

Fig. 5.3 Changes in homogamy coefficients over time

homogamy is stronger for educational groups at the two extremes of the educational distribution (junior high school and BA+).

With this heterogeneous pattern of educational homogamy in mind, Fig. 5.3 presents cohort trends in the strength of educational assortative mating from various models (Models 4 to 7). Results from Model 4 (Panel A) indicate that the relative strength of educational homogamy increased and then subsequently declined over time at all levels of educational attainment. In Panel A, we also present trends in educational hypergamy and hypogamy. Despite reasonably stable trends in the absolute levels, it appears that educational hypergamy is decreasing. In contrast, educational hypogamy is increasing across marriage cohorts, consistent with the descriptive results in Table 5.1. Similarly, we present results based on the most parsimonious representation of changes in the strength of the association in Panel B. Again, these results suggest a weakening association between husbands' and wives' educational attainment. Specifically, as compared with the initial period (men and women who married between 1975 and 1984), the association between husbands' and wives' educational attainment for the most recent cohorts declined by about 20%. This estimate is similar to findings from Fujihara and Uchikoshi (2019) who found that the association between husbands' and wives' educational attainment declined by 25% over the three decades from 1950 to 1979 (although it should be noted that the earlier study used birth cohorts rather than marriage cohorts).

5.3.2.2 Heterogeneity Within Higher Education

As we discuss in detail later, the decline in educational homogamy across marriage cohorts in Japan, especially for the highly educated (Fujihara and Uchikoshi 2019; Fukuda et al. 2021), is counter to the trends observed in other countries like the United States (Mare 1991; Schwartz and Mare 2005), where growing educational homogamy is driven by increased economic sorting in the marriage market (Schwartz and Mare 2005). Possible explanations for the distinctive pattern of declining homogamy in Japan may be the facts that wage differentials by education have not increased in recent years (Kambayashi et al. 2008; Kawaguchi and Mori 2016; Toyonaga 2021; Yamada and Kawaguchi 2015) and that highly educated women often do not remain in the labor force after childbirth (Brinton and Lee 2001; Takeuchi 2004). However, these explanations are presumably more consistent with stability in patterns of assortative mating, rather than the declining trend observed in Japan.

Another potential explanation for declining educational homogamy (especially among the highly educated) is the role of educational expansion. Educational expansion in Japan has been driven by the establishment of new private institutions, most of which are located in the lower ranks of the selectivity hierarchy (Ishida 2007). Concurrent growth in access to higher education, especially for women, suggests that the composition of college graduates in Japan is increasingly heterogeneous (Uchikoshi 2020). If the strength of educational homogamy is relatively stable among both selective college graduates and graduates of non-selective colleges, numerical growth in the latter group may decrease the overall strength of educational homogamy among college graduates if non-selective college graduates are more likely to marry partners who did not complete college.

Examining this possibility is complicated by the lack of appropriate data. One exception is a recent study by Uchikoshi (2020) who found that selective university graduates, defined as graduates of national or public universities, are more likely to marry each other than non-selective university graduates (defined as graduates of private universities). To provide further evidence regarding potential variation in the strength of educational homogamy among university graduates by college selectivity, we use data from a survey that ascertained the names of schools that both respondents and their spouses attended. These surveys were conducted from 2009 to 2011 as a part of an effort to understand the implications of income redistribution for well-being (see Oshio and Urakawa 2014 for details).[1] Because the surveys were conducted online with 20–69-year-old men and women registered with an internet panel, we created post-stratification weights to make the samples representative of the general population with respect to sex, age, and educational attainment. The total sample size was 3,083.

We classified university graduates into three groups: selective national universities, selective private universities, and non-selective universities. The selective colleges were identified using a classification scheme proposed by Yonezawa (2008), which focuses on the university's year of establishment. As Kaneko (1996) also emphasized, the selectivity of colleges in Japan is strongly correlated with the length of time since establishment. This scheme divides prestigious private universities, which were established around the year of the imperial university order in 1918, from other non-selective private universities. We classified respondents as graduates of selective universities if they graduated from the following schools:

---

[1] We thank Toshiaki Tachibanaki and Sayaka Sakoda for sharing their survey data.

Tokyo, Kyoto, Tohoku, Kyushu, Hokkaido, Osaka, Nagoya, Tokyo Institute of Technology, Tokyo Medical and Dental University, Hitotsubashi (national universities), Keio, Waseda, Meiji, Rikkyo, Hosei, Chuo, Kansai, Kwansei-Gakuin, Doshisha, Ritsumeikan, Gakushuin, and Sophia (private universities). We then constructed a category comprised of graduates of other national and public universities as our purpose here is to identify whether homogamy among selective college graduates differs from others. Table 5.4 presents the two-way cross-tabulation of spouses' education using this three-category measure of university selectivity.

Rather than seeking to identify the best fitting model, we are interested in estimating the odds ratio for educational homogamy both vertically (years of education) and horizontally (different types of university education). Figure 5.4 presents the

Table 5.4 Distribution of spouse pairing patterns (cell percent, row: males, column: females)

	JHS/HS	JC/CoT/VoC	Other Univ	Selective Private Univ	Selective National Univ	Graduate degree
JHS/HS	19.076	8.214	1.601	0.286	0.061	0.110
JC/CoT/VoC	1.505	1.036	0.473	0.003	0.000	0.085
Other Univ	13.741	15.212	10.054	0.944	0.175	0.408
Selective Private Univ	4.048	4.252	2.630	1.253	0.075	0.033
Selective National Univ	1.015	1.407	1.650	0.157	0.311	0.218
Graduate degree	1.130	2.689	4.396	0.625	0.246	0.882

Note JHS/HS-Junior high or high school, JC/CoT/VoC = junior college, college of technology, or vocational college,

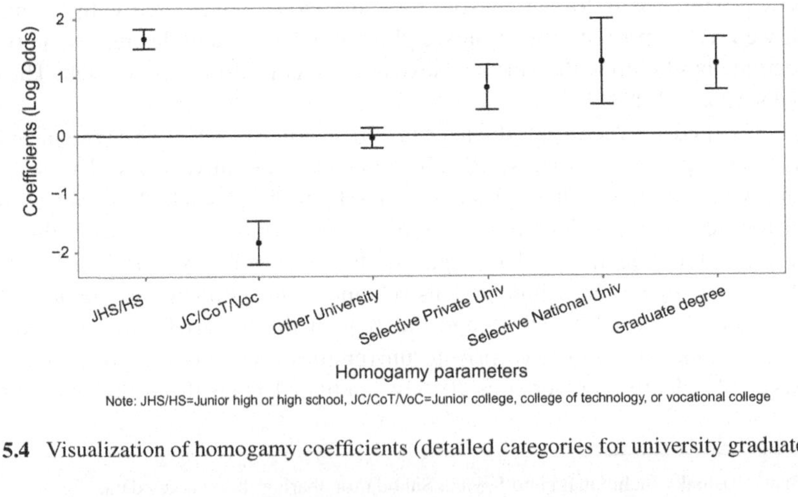

Fig. 5.4 Visualization of homogamy coefficients (detailed categories for university graduates)

odds ratios of educational homogamy by educational level.[2] Looking at university graduates, we see a clear difference in the odds of homogamy by college selectivity. Selective college graduates are more likely to marry homogamously, while the odds of homogamy among non-selective university graduates are not statistically different from those of heterogamy. More specifically, the odds ratio of homogamy is 3.46 (=exp(1.24)) among selective national university graduates and 2.25 (=exp(0.81)) among selective private university graduates. This implies that these university graduates are 3.5 times (for selective national) or 2.6 times (for selective private) more likely than other to university graduates marry spouses with the same type of education. Additionally, if these coefficients are read as the strength of homogamy, homogamy is strongest among selective national university graduates (the 95% confidence interval ranges from 0.51 to 1.98), relatively weaker among graduates of selective private universities (from 0.43 to 1.19), and weakest among graduates of other non-selective universities (from –0.22 to 0.13). Interestingly, the log odds of homogamy for those with a graduate degree are similar to those of selective college graduates. As we have already shown, the relative strength of educational homogamy is larger for lower levels of education (we combined the lowest group (junior high school) with high school graduates because of the smaller sample size) than for middle levels of education (junior colleges).

5.3.2.3 Assortative Mating by Age

We have confirmed the findings of previous studies, showing a decline in educational homogamy and an increase in educational heterogamy over time in Japan. Compared with the robust literature on educational pairing, our understanding of age assortative mating is relatively limited. Although the odds of homogamy are larger for educational attainment than for age (Qian 1998), assortative mating by age is another important reflection of both social closure and gender inequality. Empirical studies report an inverted U-shaped pattern of trends in age homogamy across countries. Age homogamy, net of distributional changes in age, increased from the beginning of the twentieth century (Atkinson and Glass 1985; Mu and Xie 2014; Ni Bhrolchain 1992; Qian 1998; Van Poppel et al.2001) before decreasing in recent years (Kolk 2015; Qian 1998; Van Poppel et al. 2001). Some scholars suggest that the rise in age heterogamy was due to an increase in cohabiting unions (Esteve et al. 2009), while other studies suggest the importance of increasing gender equality (Bozon 1991).

As discussed above, Japan is characterized by pronounced gender inequality both in the labor market and in the domestic sphere. Also, as in many European countries, cohabitation in Japan has increased, but the duration of these unions tends to be short and about half result in marriage (Raymo et al. 2009). As in other East Asian societies, it thus seems safe to say that cohabitation is not yet established as a long-term alternative to marriage in Japan (Raymo et al. 2015). In this context, we expect

[2] The model we used here is similar to Model 2 described in the Appendix.

to see limited change in age homogamy and female age hypergamy in recent years given Japan's slow progress toward gender equality in the labor market.

To examine this question, we used two publicly available administrative data sources—the vital statistics and the census. First, we used published vital statistics data to obtain the number of first marriages that were registered in a given year, by husbands' and wives' age. Next, we estimated the mid-year population of never-married men and women. Access to data on both the timing of marriage and the population at risk of marriage allows us to summarize age pairing patterns using harmonic mean models. To estimate the change in the force of attraction (marriage propensity, independent of marriage market composition) for specific age pairings, we followed Mu and Xie (2014) in focusing on the force of attraction for a given pairing divided by the sum of all forces of attraction in a particular year. Our analysis focuses on first marriages between ages 18 and 49 among never-married men and women, the population at risk of first marriage. Age homogamy is defined as marriages in which both spouses are of the same age, and robustness checks use more flexible definitions of age homogamy (i.e., marriages with age gaps of less than one or two years).

Figure 5.5 presents the trends in this measure of the force of attraction, by marriage cohort. From this figure, it is clear that age homogamy increased regardless of how it is defined, especially from 1995 to 2005. For example, age homogamy (measured as both spouses being the same age) increased by 30% from the reference year. Importantly, the increase in age homogamy is robust to different definitions (an 18% increase and an 11% increase when homogamy is defined as marriages with age gaps of less than one and two years, respectively). These results contrast with findings from

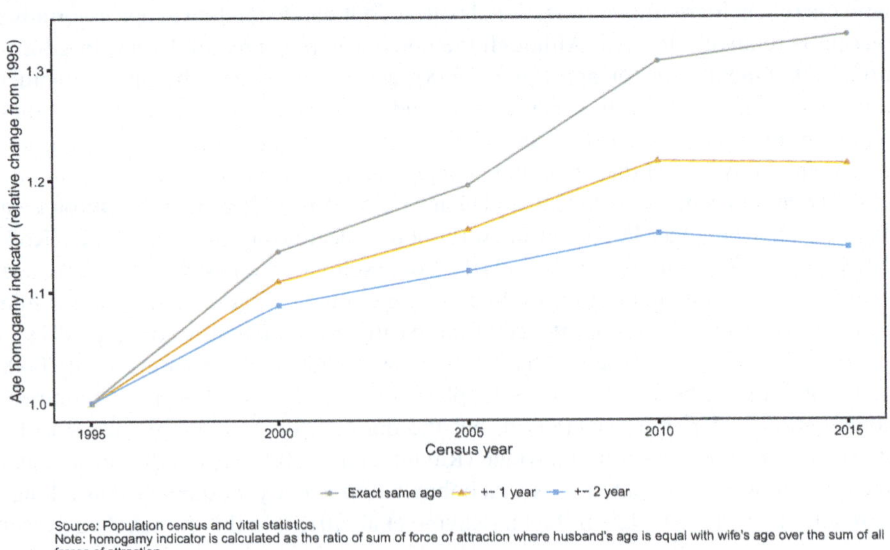

Fig. 5.5 Trends in age homogamy

previous studies demonstrating a decrease in age homogamy. The lack of information on other characteristics of husbands and wives precludes further analysis, but we see great value in evaluating and understanding potential explanations for these initial descriptive findings regarding recent trends in assortative mating with respect to age in Japan.

5.3.2.4 Status Exchange

One important question in recent studies of assortative mating is whether the rise in educational hypogamy will have substantive implications for family formation patterns. Stated differently, does the rise in educational hypogamy simply reflect the changes in educational distributions due to reversal of the gender gap in higher education, or does it also reflect changes in the nature of the "marital bargain"? One possibility is that educational hypogamy represents a new form of status exchange, as suggested by evidence that women tend to earn less than their husbands (income hypergamy) in educationally hypogamous marriages (Chudnovskaya and Kashyap 2020; Qian 2017). Given that Japanese women are disadvantaged in the labor market in terms of both employment stability and earnings, it is interesting and important to understand whether there is similar evidence of status exchange as educational hypogamy increases. To answer this question, we examined data from the Japanese Life-course Panel Study (JLPS) . By focusing on newlywed couples at the time of the survey, we can distinguish the effects of assortative mating from income dynamics after marriage. Our primary dependent variable of interest in these analyses is the odds of educational hypogamy and the main independent variable is the share of household income contributed by the respondent's spouse at the time of marriage. Our expectation is that women's share of household income is negatively associated with the likelihood of women's educational hypogamy, a pattern consistent with status exchange.

JLPS targets individual men and women, rather than households. This means that respondents provide information about their spouse's income, so it is possible (likely) that the spouse's income is reported with error. Misreporting of income would be problematic if spouses *differentially* misreport—that is, if women's misreporting is systematically different than men's misreporting. Since we view this as an empirical question, we present results for both men and women separately.

Descriptive statistics are shown in Table 5.5. Looking at the variable of primary interest, it seems that both men and women tend to report husband's income similarly, although we see greater variation for women (standard deviation of 2.4 million yen for female reports on male spouse's income versus standard deviation of 2.1 million yen for male reports on own income). A discrepancy is found for female income, however, with women tending to report their own income higher than men report their spouse's income, suggesting that men disproportionately underreport their spouse's income. The results shown below need to be interpreted carefully based on this gender difference in reporting of spouse's income.

Table 5.5 Descriptive statistics

	Male		Female		Total	
	Mean	SD	Mean	SD	Mean	SD
Educational hypogamy	0.12	0.33	0.19	0.4	0.17	0.37
Age	32.69	5.28	31.31	4.05	31.87	4.64
Respondent's annual income (million)	4.31	2.07	2.54	1.99	3.26	2.20
Spouse's annual income (million)	2.21	1.84	4.32	2.43	3.47	2.44
Respondent's income share	69.49	20.09	35.67	17.07	49.37	24.75
Respondent's standard employment	0.86	0.35	0.49	0.50	0.64	0.48
Spouse's standard employment	0.46	0.50	0.89	0.31	0.72	0.45
Observations	224		329		553	

Table 5.6 Regression results for models of educational hypogamy

	Model 1				Model 2			
	Female		Male		Female		Male	
Age	0.031	(0.036)	0.073 +	(0.042)	0.045	(0.038)	0.017	(0.046)
Respondent's annual income (million)	−0.143	(0.230)	−0.333 +	(0.189)	−0.335	(0.295)	0.071	(0.271)
Spouse's annual income (million)	−0.426**	(0.156)	−0.059	(0.307)	−0.421*	(0.178)	0.138	(0.472)
Respondent's income share	−0.015	(0.020)	0.015	(0.025)	−0.002	(0.024)	0.017	(0.034)
Respondent's standard employment	0.373	(0.395)	−0.372	(0.559)	0.324	(0.424)	−0.267	(0.626)
Spouse's standard employment	−0.354	(0.436)	1.396*	(0.652)	−0.410	(0.474)	0.900	(0.678)
Constant	0.236	(1.382)	−4.351 +	(2.378)	0.134	(1.498)	−3.145	(3.043)
Observations	329		224		274		82	
Log likelihood	−149.279		−77.848		−132.134		−50.775	
Standard errors in parentheses								

+ $p < 0.10$ * $p < 0.05$ ** $p < 0.01$ *** $p < 0.001$

Table 5.6 presents the results of logistic regression models. The first two models estimate the odds of educational hypogamy as a function of respondent's and spouse's income as well as several relevant covariates (separately by respondent's sex). For women, results indicate that husbands' income is positively associated with the relative likelihood of "traditional" spouse pairing patterns, i.e., educational homogamy and hypergamy and negatively associated with the likelihood of educational hypogamy. In contrast to husbands' income, we do not find evidence that women's own income is related to the likelihood of educational hypogamy. Further, we do not find a statistically significant relationship between the share of household income and the relative likelihood of educational hypogamy. It should be noted, however, that the coefficient is negative, suggesting that the higher the wife's income share, the lower the chance of educational hypogamy, which is similar to the findings of previous studies (Chudnovskaya and Kashyap 2020; Qian 2017). Results are similar for men, with his higher income negatively associated with the chance of marrying a woman with more education than himself.

The last two models presented in Table 5.6 evaluate the robustness of the results just described by limiting the sample to those who are at risk of educational hypogamy, namely, women who are middle- or highly educated and men who are low- or middle-educated. In general, results are similar to the previous models, especially for women. To summarize, our findings in Table 5.6 are not consistent with the findings of previous studies showing that female income hypergamy is more prevalent among educationally hypogamous couples.

5.4 Consequences for Family and Inequality

5.4.1 Economic Inequality

In contrast to the relatively comprehensive evidence on trends in educational assortative mating in Japan, we know little about its consequences, especially its implications for economic inequality. Perhaps the lack of attention to *who marries whom* in research on economic inequality reflects the fact that women's educational attainment is not strongly associated with labor force participation in Japan. Persistent gender inequality in the labor market and the highly asymmetric division of household labor have been major barriers to Japanese women (at all levels of education) seeking to balance work and family, at least until recently (Brinton and Lee 2001). As Higuchi (1991) noted, women's educational attainment has actually been negatively associated with labor force participation following marriage. The strength of this relationship is reflected in consistent support for the Douglas-Arisawa law, predicting that the secondary earner's (typically wives) economic contributions are negatively associated with the earnings provided by the primary breadwinner (typically husbands) (Kawaguchi 2002). As a result, married women's economic contributions have contributed to the mitigation of household income inequality in Japan (Abe and Oishi 2007).

Media portrayals of the rise of "power couples" raises concerns about the potential implications of trends in assortative mating for inequality and stratification in Japan (Tachibanaki and Sakoda 2013; Tsutsui 2016). However, even if highly educated women are more likely to marry highly educated men with relatively high earnings potential, which we confirmed above, the implication of these patterns for the concentration of economic resources is a question in need of careful empirical evaluation. In Japan, there is good reason to believe that an increase in educational homogamy would *not* contribute to growing inequality given that highly educated women are no less likely to leave the labor market around the time of childbirth than low-educated women (Brinton and Lee 2001) and are actually less likely to re-enter the labor market after leaving (Hirao 2005; Raymo and Lim 2011; Taki 2019). However, recent studies suggest that relationships between women's educational attainment and labor force participation may be changing. For example, some studies suggest that women's higher educational attainment is increasingly associated with labor force participation (Kohara 2008), that women's employment is becoming less dependent on husbands' income (Ohtake 2005), and that variation in married women's earnings has increased in recent years (Murakami 2001). These trends all suggest an increased role for women's earnings in shaping household income inequality (Kohara 2008). One recent study by Raymo and Iwasawa (2016) also showed that educational differences in women's labor force participation after childbearing is growing. In earlier birth cohorts (born 1940–49), employment one year after childbearing did not differ by educational attainment, but highly educated women in more recent cohorts are more likely to be employed after childbirth than their less-educated counterparts. This evidence highlights the value of examining whether changes in spouse pairing patterns by education explain part of the recent growth in economic inequality among households.

5.4.1.1 Wive's Labor Force Participation

To better understand the potential role of educational assortative mating in shaping economic inequality, we need to examine the processes through which spouse pairing patterns contribute to differences in married women's labor market outcomes and economic well-being. To this end, we first examine long-term trends in married women's employment following first childbirth, and how they differ by educational attainment, and then consider potential implications for economic inequality. To answer these questions, we used data from the SSM surveys, taking advantage of detailed information on the occupational histories of individual respondents. Unlike past research based on other nationally representative surveys (Mugiyama n.d.; Raymo and Iwasawa 2016; Senda 2002), we can thus examine the long-term consequences of childbearing for married women's labor force participation. After limiting the sample to women with at least one child and married at the time of the survey, we constructed person-year records to describe individual employment histories from one year before to ten years after first childbirth. After describing cohort trends in female employment by educational attainment, we examined whether the likelihood of employment as a standard worker differs by one's education and whether that relationship has changed in recent years. We use regression analysis to describe women's employment one year, five years, and ten years after the childbirth. To facilitate comparisons with the results of Raymo and Iwasawa's (2016) analyses, we focused on 10-year birth cohorts from 1940–49 to 1970–79.

Figure 5.6 shows women's employment status before and after childbirth for multiple birth cohorts. From these figures, it is clear that the employment rate has increased in more recent cohorts, a trend that is largely explained by the increase in non-standard work. This change in mothers' employment has been reported in a number of previous studies (e.g., Kambayashi 2017; Nagase 2004; Tsutsui 2016). Figure 5.6 also presents trends in employment status separately by mothers' educational attainment. For mothers with a high school degree or less, the trends are quite similar to the overall figure (not surprising as high school graduates comprise the majority of the sample). We see a different picture, however, for highly educated women, who are more likely to be employed in all cohorts, especially as standard employees. Although the proportion of non-standard employees has also increased, the contribution to the overall increase in employment is smaller than for high school graduates.

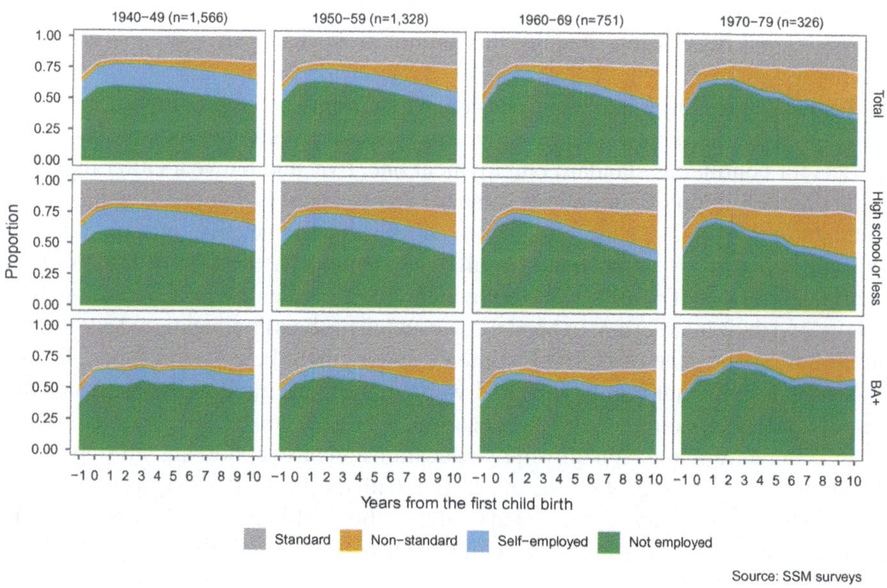

Fig. 5.6 Married women's employment status, by educational attainment

Comparing the two education groups, it seems that highly educated women have a consistently higher prevalence of standard employment, a pattern that differs from earlier studies showing a smaller educational difference in standard employment for older cohorts (Brinton and Lee 2001; Mugiyama n.d.; Raymo and Iwasawa 2016; Senda 2002). We are not able to directly investigate potential explanations for this inconsistency, but one possibility is SSM's collection of occupational information from the time respondents started working in a given job to the time they quit that job *permanently*. The SSM occupational histories may thus include short breaks when respondents were on temporary leave (e.g., parental leave), but still in the labor force. In contrast, the surveys used in previous studies have presumably treated mothers on temporary leave as not employed. This distinction is critical when we think about educational differences in women's employment after childbearing because highly educated women in older cohorts had few occupational opportunities to develop and use professional skills. As a result, these women typically worked as teachers in primary or secondary education, one of the first occupations to implement a parental leave policy for mothers with small children (Tanaka and Nishimura 1986).

Evidence of this historical context can also be seen in Panel A of Table 5.7, which presents the distribution of employment types and occupation, by mother's educational attainment, at the time of birth for women in the oldest cohort (born in 1940–49). Nearly 35% of university-educated women worked in standard employment at the time of first childbirth, a figure that is significantly higher than that of high school graduates (20%) or junior college graduates (22%). The higher proportion of standard employment among college-educated women in this cohort is explained by their occupation, with Panel B of Table 5.7 showing that a majority worked in a professional occupation. Among this group, more than half (60%) worked as a teacher.

To further explore educational differences in women's employment after childbirth, we used the person-year data described above to estimate logistic regression models for continuous standard employment one, five, and ten years after the childbirth. Our definition of continuous standard employment is that one is working as a standard employee both one year before childbirth and one, five, or ten years after childbirth. Table 5.8 presents the results of two models, the first of which examines the correlates of continuous employment, focusing on mother's education, birth cohort, and their interaction. In Model 2, we added husband's educational attainment, an important correlate of married women's employment behavior in Japan. Results of Model 1 show that highly educated women are more likely to work in standard employment than women with less education, but there is no evidence of change over time—interactions between university education and birth cohort are not statistically different from zero (except for the interaction with the 1970–1979 birth cohort). This is consistent with the aforementioned evidence that highly educated women in older cohorts were concentrated in the teaching profession and thus eligible for child care leave (Tanaka 1997).

Table 5.7 Distribution of occupation, by education at the time of childbirth

Panel A

	High school or less	Junior college	BA +	Total
Standard	271	31	24	326
Row percent	20.1	21.5	33.8	20.8
Non-standard	49	3	1	53
Row percent	3.6	2.1	1.4	3.4
Self-employed	233	24	9	266
Row percent	17.2	16.7	12.7	17
Not employed	798	86	37	921
Row percent	59.1	59.7	52.1	58.8
Total	1351	144	71	1566
Row percent	100	100	100	100

Panel B

	High school or less	Junior college	BA +	Total
Proffetional/Managerial	39	11	11	61
Row percent	2.9	7.6	15.5	3.9
Teaching	4	3	16	23
Row percent	0.3	2.1	22.5	1.5
Clerical/Sales	238	33	5	276
Row percent	17.6	22.9	7	17.6
Manual	200	9	1	210
Row percent	14.8	6.2	1.4	13.4
Farming	70	1	1	72
Row percent	5.2	0.7	1.4	4.6
Not employed	800	87	37	924
Row percent	59.2	60.4	52.1	59
Total	1351	144	71	1566
Row percent	100	100	100	100

Results of Model 2 provide no evidence that husbands' educational attainment is significantly associated with women's labor force participation after childbearing. While earlier studies consistently demonstrated that husbands' educational attainment and earnings were negatively related to married women's employment after childbearing (Higuchi 1991; Higuchi et al. 2016; Mincer 1985; Nagase 1999; Wakisaka and Tomita 2001), more recent work suggests that this relationship has weakened or disappeared (Kohara 2008; Raymo and Lim 2011; Waldfogel et al. 1999). Importantly, the fact that we do not find any evidence of a link between husbands' education and wives' post-childbirth employment, combined with highly educated mothers' more consistent employment, suggests that growth

Table 5.8 Regression results for models of standard employment at one, five, and ten years after the first childbirth

	Model 1						Model 2					
	1 year		5 years		10 years		1 year		5 years		10 years	
Junior college	0.205 +	(0.111)	0.201 +	(0.122)	0.256*	(0.129)	0.256*	(0.119)	0.254 +	(0.131)	0.293*	(0.139)
BA +	0.990***	(0.264)	1.034***	(0.276)	1.209***	(0.278)	1.048***	(0.280)	1.088***	(0.295)	1.233***	(0.298)
Survey year: 1995	0.196	(0.207)	0.195	(0.218)	0.239	(0.234)	0.245	(0.215)	0.265	(0.227)	0.324	(0.246)
Survey year: 2005	0.062	(0.190)	−0.070	(0.201)	−0.042	(0.217)	0.122	(0.197)	0.006	(0.210)	0.058	(0.228)
Survey year: 2015	−0.193	(0.192)	−0.320	(0.204)	−0.242	(0.219)	−0.121	(0.199)	−0.225	(0.213)	−0.122	(0.231)
Birth cohort: 1950–59	0.176 +	(0.106)	0.192 +	(0.116)	0.149	(0.124)	0.179 +	(0.107)	0.204 +	(0.117)	0.158	(0.125)
Birth cohort: 1960–69	0.174	(0.134)	0.168	(0.148)	0.115	(0.158)	0.180	(0.135)	0.185	(0.149)	0.129	(0.160)
Birth cohort: 1970–79	0.377*	(0.179)	0.330 +	(0.199)	0.302	(0.211)	0.370*	(0.181)	0.336 +	(0.202)	0.309	(0.214)
Birth cohort: 1950–59 # BA +	−0.210	(0.339)	−0.163	(0.354)	−0.172	(0.357)	−0.182	(0.342)	−0.138	(0.358)	−0.146	(0.362)
Birth cohort: 1960–69 # BA +	−0.223	(0.361)	0.024	(0.373)	−0.144	(0.383)	−0.189	(0.364)	0.049	(0.377)	−0.125	(0.387)
Birth cohort: 1970–79 # BA +	−0.480	(0.438)	−0.758	(0.491)	−1.059*	(0.525)	−0.503	(0.448)	−0.830	(0.509)	−1.167*	(0.552)
Spouse: Junior college							−0.096	(0.169)	−0.277	(0.196)	−0.311	(0.213)
Spouse: BA +							−0.138	(0.110)	−0.140	(0.120)	−0.094	(0.127)
Constant	−1.693***	(0.172)	−1.850***	(0.181)	−2.070***	(0.196)	−1.729***	(0.180)	−1.900***	(0.191)	−2.150***	(0.209)
Observations	3968		3968		3968		3906		3906		3906	
Log likelihood	−1871.396		−1650.960		−1500.937		−1838.305		−1617.903		−1467.189	
Standard errors in parentheses												

+ $p < 0.10$ * $p < 0.05$ ** $p < 0.01$ *** $p < 0.001$

in the number of educationally homogamous couples may contribute to increasing income inequality.

5.4.1.2 Implications for Income Inequality

In their review article, McCall and Percheski (2010) discussed three ways in which family may contribute to increasing inequality: changes in family structure, women's increased access to the labor market, and an increase in educational assortative mating. In terms of changing family structure, growth in single-parent households is seen as particularly important (McLanahan and Percheski 2008), although its implications for inequality vary across countries depending on the generosity of public income support (Gornick and Jäntti 2012). In contrast, the rise in women's labor force participation, which is often negatively correlated with husbands' income, has had the effect of equalizing household income (Cancian and Reed 1998; Kollmeyer 2013; Treas 1987).

Results relevant to the third mechanism, and our primary focus here, are inconsistent. A null association between trends in educational assortative mating and growing household inequality has been found in Britain (Breen and Salazar 2010), the United States (Greenwood et al. 2014; Hryshko et al. 2017), and four Western countries including the United States (Eika et al. 2019). Other studies have found that the rise in educational assortative mating contributed to a reduction in economic inequality in some countries like China (Hu and Qian 2015) and the United States (Breen and Salazar 2011). Yet other research has linked the rise in educational assortative mating to increasing inequality in countries like Denmark where the rise in women's labor force participation has been large (Breen and Andersen 2012; Esping-Andersen 2007). This reflects the facts that highly educated women are more likely to have full-time employment and increasingly tend to marry men with similarly high levels of education.

Efforts to synthesize this large body of inconsistent evidence suggest that the impact of educational assortative mating on economic inequality depends on the degree to which women's education is associated with employment (Herzberg-Druker and Stier 2019; Schwartz 2013). We believe that another relevant factor is the degree of gender inequality in society and its implications for highly educated women's incentives to marry (or enter a union, more generally). According to the "shifting economic foundations of marriage" thesis (Oppenheimer 1988; Oppenheimer et al. 1997; Sweeney 2002), the opportunity costs of marriage and childbearing for highly educated women are higher in societies where women are normatively expected and structurally encouraged to concentrate on domestic labor rather than market labor following marriage and childbearing. However, it is important to recognize how changes in men's and women's economic position in the labor market increase the value of women's economic contributions to the family and thus reduce incentives for gender specialization within marriage.

In light of this emphasis on the need to consider both spouse pairing patterns and the role of women's employment subsequent to family formation, it is important to recognize that the link between women's educational attainment and employment in Japan is relatively weak (Brinton and Lee 2001). While there is some evidence of a growing educational gradient in stable employment in Japan (Mugiyama n.d.; Raymo and Iwasawa 2016; Senda 2002), our analyses in the last section show that it is not strong. Nevertheless, it is the case that growth in non-standard employment is more prevalent among less-educated women, suggesting the value of examining relationships between educational assortative mating and income inequality with a focus on women's employment patterns after childbearing.

As in the last section, we used SSM surveys from 1985 to 2015. We limited our analytical samples to married women with children who reported their occupational history. The main outcome is the sum of husbands' and wives' income, which we define as household income. We constructed the variable on educational assortative mating by aggregating educational attainment into three groups, low (junior high and high schools), middle (junior colleges and vocational schools), and high (BA +), and four pairing types; homogamy (low or middle educated), homogamy (highly educated), hypogamy (women marrying down), and hypergamy (women marrying up). Based on prior research, we posit that highly educated women in educationally homogamous marriages are less likely to continue their employment after childbirth, thus mitigating the potential link between increasing educational homogamy and income inequality. Based on the information about women's occupational history, we constructed four employment patterns; (1) continuous standard employment while balancing work and family (career *and* family), (2) continuous standard employment while prioritizing work (career *then* family), (3) continuous employment other than standard employment (job), and (4) career interruption (interruption). This classification is inspired by Antecol (2015) who examined employment patterns among women who graduated from elite colleges in the United States. Continuous employment is defined as those with no spells out of the labor market for the first 10 years after initially entering the labor market. "And family" means that they experienced childbirth during these first 10 years. "Then family" refers to women who had a childbirth 10 or more years after initial labor market entry. "Interruption" means that there is at least one year out of the labor market during the first 10 years. To consider the critical difference between working as a regular or non-standard employee in the Japanese labor market, we also distinguished continuous employment in standard employment and other types of work. We call the former "career" employment and the latter a "job."

Tables 5.9 and 5.10 show the distribution of educational assortative mating and employment patterns at the four points in time (1985, 1995, 2005, 2015). As expected, there is a clear increase in both educational homogamy among highly educated men and women (4.4% to 12.9%) and women's educational hypogamy (3.7–14.7%). The trends in employment patterns also show that the proportion of women who continue their employment as a standard worker has increased gradually. Although the increase in female labor force participation in Japan is explained by growth in non-standard employment (Kambayashi 2017; Nagase 2004; Tsutsui 2016), women in non-standard work tend to experience labor force interruption around first childbirth.

Table 5.9 Distribution of educational assortative mating

	Homogamy (low and middle)	Homogamy(high)	Hypogamy	Hypergamy	Total
1985	517	32	27	155	731
Row percent	70.7	4.4	3.7	21.2	100.0
1995	562	35	29	180	806
Row percent	69.7	4.3	3.6	22.3	100.0
2005	1208	122	110	490	1930
Row percent	62.6	6.3	5.7	25.4	100.0
2015	864	272	310	662	2108
Row percent	41.0	12.9	14.7	31.4	100.0
Total	3151	461	476	1487	5575
Row percent	56.5	8.3	8.5	26.7	100.0

Table 5.10 Distribution of employment patterns among married women with children

	Career and family	Career then family	Job	Interruption	Total
1985	70	56	188	417	731
Row percent	9.6	7.7	25.7	57.0	100.0
1995	96	72	170	468	806
Row percent	11.9	8.9	21.1	58.1	100.0
2005	197	188	320	1225	1930
Row percent	10.2	9.7	16.6	63.5	100.0
2015	232	235	283	1358	2108
Row percent	11.0	11.1	13.4	64.4	100.0
Total	595	551	961	3468	5575
Row percent	10.7	9.9	17.2	62.2	100.0

Table 5.11 Two-way cross-tabulation of educational assortative mating and employment patterns

	Career and family	Career then family	Job	Interruption	Total
Homogamy (low and middle)	298	323	634	1896	3151
Row percent	9.5	10.3	20.1	60.2	100.0
Homogamy(high)	90	30	53	288	461
Row percent	19.5	6.5	11.5	62.5	100.0
Hypogamy	71	31	78	296	476
Row percent	14.9	6.5	16.4	62.2	100.0
Hypergamy	136	167	196	988	1487
Row percent	9.1	11.2	13.2	66.4	100.0
Total	595	551	961	3468	5575
Row percent	10.7	9.9	17.2	62.2	100.0

Importantly, Table 5.11 shows that educational assortative mating and employment patterns are correlated. First, we see that treating highly educated homogamous couples as "power couples" (Compton and Pollak 2007; Costa and Kahn 2000; Tachibanaki and Sakoda 2013) is somewhat misleading. Although women in these couples are more likely to pursue careers in standard employment (19.5%) than women in other types of educational pairing, the percent in the interruption subgroup is large (62.5%). This is not surprising in light of earlier studies demonstrating that the connection between women's education and employment after childbearing is relatively weak (Brinton and Lee 2001), that women's employment is negatively associated with husbands' earnings, and that highly educated women's employment behavior is quite heterogeneous (Hirao 1999; Raymo and Lim 2011). As such, the share of what we might call "power couples" is limited in this sample (Uchikoshi 2018).

Second, we see that women in educationally hypogamous marriages are more likely to pursue careers as standard workers while experiencing childbirth during the first 10 years of employment. The implications of the rise in educational hypogamy couples are thus unclear. On the one hand, to the extent that mothers' career trajectories are associated with higher individual income, the increase in educationally hypogamous couples may contribute to growing household income inequality. On the other hand, these couples might be also characterized by husbands with lower earnings than men in highly educated homogamous couples. To explore further, we investigated the income distribution across educational pairing and employment patterns.

Mean income by educational assortative mating and employment patterns are presented in Tables 5.12 and 5.13. Husbands' and wives' incomes are highest in highly educated homogamous couples. Wives' income is also higher in educationally hypogamous couples, perhaps reflecting a higher prevalence of career-oriented women in this group, while it is also the case that their husbands' income is lower than that of men in highly educated homogamous and hypergamous unions. These results suggest that the rise in educational homogamy among the highly educated and in educationally hypogamous couples may be related to income inequality in complicated ways. On the one hand, the rise of highly educated homogamous couples may contribute to the growing income inequality if women in these couples pursue their careers similar to women in hypogamous couples. However, if a sizable proportion of these women do not pursue careers, we would not expect an increase in highly educated homogamous couples to contribute to growing inequality. On the other hand, there is good reason to believe that the rise of educationally hypogamous couples may contribute to income inequality as women in these couples are more likely to pursue a higher-earning, career-oriented life-course. However, these couples are also characterized by lower-earning husbands, thus we can also expect that the contribution of growth in these pairings to inequality is limited. To examine the potential role of educational assortative mating for income inequality, we test these possibilities using a counterfactual approach.

Table 5.14 presents the results of our counterfactual analysis. First, we see from the observed Theil index that income inequality has increased over time (from 0.14

Table 5.12 Wife's, husband's, and couple's income, by educational assortative mating

	Homogamy (low and middle)		Homogamy(high)		Hypogamy		Hypergamy		Total	
	Mean	SD	Mean	SD	Mean	SD	Mean	SD	Mean	SD
Husband's income	364.9	274.2	602.5	483.0	411.3	290.8	499.4	363.4	424.4	332.1
Wife's income	115.0	149.3	184.8	248.1	165.3	188.3	123.7	182.1	127.4	173.3
Household income	479.9	327.5	787.3	537.8	576.5	348.0	623.1	424.3	551.8	389.5
N	5575									

Table 5.13 Wife's, husband's, and couple's income by wife's employment patterns

	Career and family		Career then family		Job		Interruption		Total	
	Mean	SD	Mean	SD	Mean	SD	Mean	SD	Mean	SD
Husband's income	409.7	272.7	418.4	311.0	369.4	292.3	443.1	352.7	424.4	332.1
Wife's income	260.5	227.3	171.5	224.0	121.1	155.6	99.3	143.5	127.4	173.3
Household income	670.3	393.3	589.9	423.4	490.5	351.9	542.4	388.5	551.8	389.5
N	5575									

Table 5.14 Observed and counterfactual household income inequality measured as Theil index

	Year	Value
Observed		
	1985	0.14
	1995	0.16
	2005	0.23
	2015	0.20
Counterfactual		
Distribution of educational assortative mating constant	2015	0.20
Ratio	2015	1.03
Distribution of employment status constant	2015	0.21
Ratio	2015	1.05

to 0.20). We construct a counterfactual Theil index for 2015 by keeping the distribution of educational pairings constant at their 1985 values. This allows us to answer the question "What would household income inequality be in 2015 if patterns of spouse pairing with respect to educational attainment had counterfactually remained unchanged since 1985?" The ratio of counterfactual inequality index to the observed index is greater than 1.0, suggesting that changing patterns of spouse pairing have acted to lower household income inequality. This may appear surprising in light of the increase in homogamy among highly educated men and women, but previous studies have found similar results in other countries (Breen and Salazar 2011; Hu and Qian 2015). In the Japanese context, this finding makes good sense in light of the evidence we have just presented showing that highly educated women in homogamous marriages are quite heterogeneous, with many in continuous career employment and many with interrupted employment. This suggests that efforts to understand trends in household income inequality should focus primarily not on trends in educational assortative mating, but rather on women's changing employment patterns. We test this claim by constructing a counterfactual Theil index where the distribution of wives' employment patterns is held constant over time. The ratio of this counterfactual index to the observed value suggests that household income inequality would be much higher if there had been no change in women's employment patterns. These results suggest that both assortative mating and employment patterns have contributed to lower income inequality, especially the latter.

Looking at the levels of income inequality within groups helps in the interpretation of overall patterns. For example, because educationally hypergamous couples are characterized by relatively high-income inequality while hypogamous couples have lower levels of inequality, the relative increase in the prevalence of the latter group contributed to lower overall levels of income inequality. Moreover, we can see that career interruption is both associated with higher levels of inequality and more prevalent among educationally hypergamous couples. Meanwhile, career employment is

Table 5.15 Inequality within educational assortative mating types

EAM	Homogamy (low and middle)	Homogamy (high)	Hypogamy	Hypergamy
	0.196	0.189	0.15	0.199
Employment	Career and family	Career then family	Job	Interruption
	0.166	0.188	0.205	0.216

associated with lower levels of inequality and is more prevalent among educationally hypogamous couples. To summarize, the rise in homogamy among the highly educated and in educational hypogamy has not led to growing inequality because: (1) women's employment patterns in the former group are heterogeneous and (2) the income of husbands in educationally hypogamous couples is relatively low (Table 5.15).

5.4.2 *Family Formation Outcomes*

5.4.2.1 Fertility

One potential demographic consequence of the rise in educational hypogamy is a reduction in fertility (Nomes and van Bavel 2016; Osiewalska 2017). Specifically, previous studies predict that women in educationally hypogamous unions tend to delay childbirth or have fewer children for a variety of reasons. As discussed above, one theory posits that educational hypogamy is associated with later childbearing and fewer children because women in such unions are likely to contribute significantly to household income and thus face larger opportunity costs of childbearing (Klesment and van Bavel 2017; van Bavel and Klesment 2017).[3] Another potential mechanism is that women with lower childbearing preferences are more likely to select into educationally hypogamous marriages. Finally, it is also possible that the negative association between educational hypogamy and childbearing simply reflects the tendency for women in these unions tends to be highly educated (and thus have lower fertility for reasons related primarily to educational attainment rather than educational pairing).

While these scenarios lead us to expect lower fertility among educationally hypogamous unions, there is also good reason to believe that the opposite might be true for some subgroups. For example, consider women in the highest educational group (BA + in this study) who can either marry hypogamously or homogamously with respect to educational attainment. For these women, educational hypogamy may be

[3] It could also be the case that wives' job stability may also matter for childbearing decisions (Bueno and García-Román 2021). In contexts characterized by greater economic (employment) uncertainty, we may see a smaller difference in fertility levels between educationally hypogamous couples and other couples.

positively related to the number of children if it is the case that highly educated homogamous couples are more likely to prioritize the quality of children over the quantity (Becker 1991; Nitsche et al. 2018). In this case, we should see a positive association between educational hypogamy and the fertility outcomes of highly educated women.

To examine these possibilities, we used JLPS data to evaluate both the desired number of children and the actual number of children for women. Since fertility desires were asked only in Waves 1, 4, 6, and 8, we limit our analytical sample to respondents in these waves to facilitate comparison of results for the two different fertility outcomes.

Table 5.16 presents the results of Poisson regression models with individual random effects. First, we find no evidence of a negative association between educational hypogamy and the desired number of children. The coefficient for educational hypogamy is close to zero in Model 1 and it did not change after adding several potentially relevant covariates in Model 2. However, models that estimate the actual number of children born provide a somewhat different picture. Although we do not find a statistically significant association between educational pairing and fertility in Model 1, women's educational hypogamy is positively associated with the number of children (at the 10% level of significance) in Model 2. We speculate that this pattern may reflect a prioritization of child quality over quantity in marriages involving highly educated men (or perhaps a more gender-equal division of domestic labor in educationally hypogamous couples). The alternative to hypogamy for highly educated women is, of course, homogamous marriage to a highly educated man and, as we have seen above, women in the middle educational category are more likely to marry hypergamously than homogamously. We consider this speculative interpretation focused on the quantity–quality tradeoff in further detail below.

5.4.2.2 Divorce

Couples' similarity with respect to a wide range of characteristics, including indicators of socioeconomic status, has long been recognized as an important correlate of relationship stability. In this context, it is important to emphasize that status hypogamy (women marrying down) not only involves economic and cultural dissimilarity, but has also been viewed as a non-normative pairing pattern. When social norms emphasize men's role as the primary breadwinner, women who outearn their husbands report lower levels of marital satisfaction and psychological well-being (Hornung and McCullough 1981) which may contribute to their higher likelihood of experiencing divorce (Heckert et al. 1998; Tzeng 1992).

In wealthy countries where women now outnumber men in college attendance and completion (DiPrete and Buchman 2012), some scholars talk of a reversal of gender inequality and highlight the increasing prevalence of educational hypogamy (Esteve et al. 2012, 2016; van Bavel 2012; van Bavel et al. 2018). As this type of couple becomes increasingly common, some scholars argue that the nature of

Table 5.16 Regression results estimating the actual and desired number of children

	Model 1		Model 2		Model 3		Model 4	
	Desired number of children		Desired number of children		Number of children		Number of children	
Age	−0.010***	(0.002)	−0.009***	(0.002)	0.032***	(0.003)	0.034***	(0.003)
Number of children	0.139***	(0.009)	0.138***	(0.009)				
Junior college	0.012	(0.022)	0.013	(0.024)	−0.068*	(0.034)	−0.057	(0.034)
BA +	−0.030	(0.026)	−0.032	(0.029)	−0.190***	(0.041)	−0.168***	(0.042)
Educational hypogamy	0.003	(0.024)	−0.003	(0.026)	0.055	(0.038)	0.072 +	(0.038)
Female partner's standard employment			0.029	(0.033)			0.040	(0.047)
Respondent's annual income (million)			−0.005	(0.013)			−0.012	(0.019)
Spouse's annual income (million)			−0.003	(0.005)			−0.002	(0.006)
Respondent's income share			0.001	(0.001)			−0.007***	(0.002)
Desired number of children					0.184***	(0.012)	0.183***	(0.012)
Constant	1.031***	(0.067)	1.007***	(0.074)	−1.189***	(0.108)	−1.091***	(0.112)
Observations	4952		4453		4952		4453	
Standard errors in parentheses								

+ $p < 0.10$ * $p < 0.05$ ** $p < 0.01$ *** $p < 0.001$

these relationships and their association with marital quality and marital stability has changed in the United States (Schwartz and Han 2014).

Research focusing on the implications of the rise in educational hypogamy for divorce and other family outcomes is limited (van Bavel et al. 2018). The relative lack of attention to these pairings may reflect the fact that they have been relatively few in number, especially in societies like Japan where women's educational hypergamy has been far more normatively desirable and numerically common. However, unmarried men and women increasingly indicate a desire to have a dual-earner household, and attitudes regarding women's economic contributions to the family are changing (IPSS 2017). If the rise in educational hypogamy described above reflects these normative and behavioral changes regarding husbands' and wives' roles, it is possible that the relative risk of divorce for educationally hypogamous couples may have decreased. Indeed, one recent study finds support for this speculation (Uchikoshi 2019). However, we see value in re-examining this question to address the fact that the earlier study also included women who were not able to marry hypogamously (i.e., those in the lowest educational group). To provide a more meaningful comparison with what we have done in previous sections, we estimate models of divorce among women for whom educational hypogamy is possible.

To better understand the potential mechanisms linking educational hypogamy and divorce, it is useful to examine whether and how the relationship has changed over time. To this end, we estimated discrete-time hazard models via logistic regression using JLPS data. The outcome in this analysis is the experience of divorce between Waves 2 to 9 of the survey. In addition to educational pairing, we examine two posited explanations for the positive association between educational hypogamy and the risk of divorce found in previous studies. One is spouses' shared interests, measured as sharing a hobby, sharing values, and having a similar upbringing/social background. We also consider the role of women's income in both absolute and relative terms, where the latter is defined as the share of wives' income in total spousal income. Descriptive statistics are shown in Table 5.17.

The results of discrete-time logistic regression models are shown in Table 5.18. In Model 1, we examine the association between educational pairing and the risk of divorce, finding that educational hypogamy is associated with a risk of divorce that is 3.7 times (=exp(1.3)) higher than for homogamy or hypergamy. Model 2 includes the measures of spousal similarity, showing that shared values are associated with a lower risk of divorce (statistically significant at the 10% threshold), but we do not find evidence that shared hobbies and similar backgrounds are related to divorce. Specifically, those who reported sharing values with their spouse are 2.68 times (=1/exp(–0.987)) less likely to divorce than those who do not share values. Most importantly, the inclusion of these shared traits results in some attenuation of the positive coefficient for educational hypogamy (from 1.31 to 1.17). Results of Model 3 show that both wives' and husbands' earnings are negatively associated with the risk of divorce, but wives' share of total spousal income is positively associated with divorce. After adding the income variables, Model 4 shows that the coefficient for shared values is no longer significantly different from zero. Importantly, educational hypogamy is still positively associated with the risk of divorce although the size of

Table 5.17 Descriptive statistics

	Mean	SD	Min	Max
Divorce	0.00	0.06	0.00	1.00
Age	37.70	5.09	24.00	49.00
BA +	0.32	0.47	0.00	1.00
Educational hypogamy	0.29	0.45	0.00	1.00
Shared traits: hobby	0.27	0.44	0.00	1.00
Shared traits: values	0.43	0.50	0.00	1.00
Shared traits: family background	0.15	0.36	0.00	1.00
Respondent's annual income (million)	1.45	1.77	0.00	25.00
Spouse's annual income (million)	5.36	2.98	0.00	25.00
Respondent's income share	20.19	20.73	0.00	100.00
Observations	5313			

Table 5.18 Regression results estimating the risk of divorce

	Model 1		Model 2		Model 3		Model 4	
Age	0.088*	(0.045)	0.089*	(0.044)	0.033	(0.045)	0.038	(0.046)
BA +	−0.038	(0.485)	0.076	(0.487)	0.588	(0.556)	0.624	(0.558)
Educational hypogamy	1.307**	(0.437)	1.171**	(0.442)	1.126*	(0.497)	1.070*	(0.503)
Shared traits: hobby			−0.020	(0.566)			0.437	(0.644)
Shared traits: values			−0.987 +	(0.569)			−0.929	(0.640)
Shared traits: family background			−1.003	(1.035)			−0.231	(1.135)
Respondent's annual income (million)					−0.742*	(0.338)	−0.847*	(0.358)
Spouse's annual income (million)					−0.710*	(0.337)	−0.623 +	(0.341)
Respondent's income share					0.048**	(0.018)	0.054**	(0.019)
Constant	−9.452***	(1.848)	−9.060***	(1.837)	−6.020**	(1.985)	−6.336**	(2.038)
Observations	5313		5313		5313		5313	
Log likelihood	−136.299		−133.360		−80.165		−78.754	
Standard errors in parentheses								

+ p < 0.10 * p < 0.05 ** p < 0.01 *** p < 0.001

the coefficient was further reduced (from 1.17 to 1.07), suggesting that part of the association between educational hypogamy and divorce is explained by hypogamous women's larger income share.

5.4.2.3 Gender Division of Labor

Another potential implication of the rise in educational hypogamy is its effect on gender equality within households. A small number of recent studies have asked whether attention to spouse pairing patterns can improve our understanding of the gender division of labor (Bonke and Esping-Andersen 2011; Miller 2020). Intuition from theories of bargaining power within marriage suggests that men are likely to spend more time on domestic work when married to a woman who has higher educational attainment than themselves (Miller 2020). The question we address is whether educational hypogamy is associated with a more equal gender division of labor (relative to homogamy and hypergamy), net of established correlates of power dynamics within a marriage. To the best of our knowledge, there is only one previous study that examines this question (Miller 2020).

To answer this question in the Japanese context, we use JLPS data. In most waves (except for Waves 2 and 4) respondents were asked about time spent on domestic work both by themselves and by their spouse. We calculated the share of husbands' and wives' housework using four measures of housework—preparing meals, cleaning, shopping, and doing laundry (see Fuwa 2020 for a similar approach). The outcome variable is husbands' share of housework, which averages about one-fourth in the JLPS data. Many studies have documented men's limited housework contributions in Japan (Ishii-Kuntz et al. 2004; Tsutsui 2016; Tsuya et al. 2000, 2012), but our interest here is to understand whether men's contribution differs by spouses' educational pairing.

As we already noted in the analysis of status exchange, JLPS targets individual men and women, rather than households. Also, there is good reason to treat men's and women's reports on housework responsibilities separately, as a number of studies found that couples do not agree on how much housework each partner does (Geist 2010). Specifically, women tend to report fewer hours of housework by their husband's relative to the housework hours reported by men themselves (Kamo 2000). In light of this gender difference in reporting of housework, we estimate results for men and women separately.

To make our results comparable with the earlier study by Miller (2020), we stratified our sample by respondents' educational attainment. The rationale for this is the fact that counterfactual possibilities differ by educational attainment. For example, men with the highest education (BA or more) cannot, by definition, marry a woman with more education than themselves. We, therefore, estimate models for six subsamples; low educated (high school or less), middle educated (junior colleges), and highly educated (BA +) for men and women. Descriptive statistics are shown in Table 5.19.

Table 5.19 Descriptive statistics

	Male				Female				Total			
	Mean	SD	Min	Max	Mean	SD	Min	Max	Mean	SD	Min	Max
Respondent's share of housework	0.30	0.11	0.14	0.82	0.73	0.10	0.14	0.86	0.55	0.23	0.14	0.86
Partner's share of housework	0.70	0.11	0.18	0.86	0.27	0.10	0.14	0.86	0.45	0.23	0.14	0.86
Age	38.44	5.33	21.00	49.00	38.13	5.46	21.00	49.00	38.26	5.41	21.00	49.00
High school or less	0.33	0.47	0.00	1.00	0.31	0.46	0.00	1.00	0.32	0.47	0.00	1.00
Junior college	0.22	0.41	0.00	1.00	0.45	0.50	0.00	1.00	0.36	0.48	0.00	1.00
BA +	0.45	0.50	0.00	1.00	0.23	0.42	0.00	1.00	0.33	0.47	0.00	1.00
Educational homogamy	0.44	0.50	0.00	1.00	0.44	0.50	0.00	1.00	0.44	0.50	0.00	1.00
Educational hypergamy	0.31	0.46	0.00	1.00	0.38	0.49	0.00	1.00	0.35	0.48	0.00	1.00
Educational hypogamy	0.25	0.43	0.00	1.00	0.18	0.39	0.00	1.00	0.21	0.41	0.00	1.00
Respondent's standard employment	0.87	0.34	0.00	1.00	0.25	0.43	0.00	1.00	0.51	0.50	0.00	1.00
Partner's standard employment	0.26	0.44	0.00	1.00	0.86	0.34	0.00	1.00	0.61	0.49	0.00	1.00
Respondent's income share	80.14	19.91	0.00	100.00	21.55	20.33	0.00	100.00	46.17	35.25	0.00	100.00
Observations	15,413											

Table 5.20 presents the results of the random-effects regression models. For men, results indicate that educational assortative mating plays a limited role in their contribution to housework. This is consistent with previous work showing little variation in men's housework (Tsuya et al. 2000). Among low- and middle-educated men, we do not find evidence that educational pairing is related to the share of housework, but educational hypergamy is negatively correlated with highly educated men's contribution to housework. This is consistent with expectations that educationally hypergamous marriages are more likely to be characterized by gender-based specialization in employment and family (Miller 2020).

Results for women are somewhat different than those for men. First, we found no statistically meaningful association between educational pairing and husband's share of housework among low-educated women. Second, although statistically significant only at the 10% level, it is interesting that husbands' share of housework is lower for both educational hypergamy and hypogamy among middle-educated women. The negative association between educational hypogamy and male partners' share of housework is somewhat surprising in light of the theoretical expectations articulated above. This unexpected pattern is also visible for highly educated women—husbands' share of housework is lower among educationally hypogamous couples. Importantly, these relationships remain after controlling for other potentially relevant covariates like wives' share of household income and employment status.

These results are not consistent with the equalization thesis positing that men who marry women more highly educated than themselves are more likely to share housework responsibilities than those who marry homogamously. Indeed, our results are more consistent with expectations of the "doing gender" or gender display perspective, in which women who outearn their husbands tend to spend relatively more time on housework to compensate for the deviation from gender norms (West and Zimmermann 1987). Of course, our focus is on educational pairing rather than relative income, but our findings are interesting and merit further investigation.

5.4.3 Diverging Destinies

As the economic foundations of marriage have shifted toward an increasing prioritization of both spouses' economic contributions (Sweeney 2002), we expect to see more pronounced positive assortative mating by socioeconomic resources (Schwartz and Mare 2005). In this context, a small number of studies have asked whether associated trends in educational assortative mating may have implications for intergenerational mobility. While prior research has often focused on the intragenerational implications of educational assortative mating for income inequality, it is reasonable to posit that the increase in educational assortative mating in the United States and elsewhere may have an impact on children's educational achievement and subsequent life outcomes. McLanahan (2004) coined the term "diverging destinies" to discuss the possibility that growing socioeconomic differences in family behavior contribute to the reproduction of inequality via their implications for the resources

Table 5.20 Regression for models of the share of housework

	Own share of housework						Partner's share of housework					
	Male (low)		Male (middle)		Male (high)		Female (low)		Female (middle)		Female (high)	
Age	0.001	(0.000)	0.000	(0.001)	0.001	(0.000)	0.000	(0.000)	0.000	(0.000)	−0.001*	(0.000)
Educational hypogamy	−0.001	(0.006)	0.014	(0.010)					−0.011 +	(0.006)	−0.020*	(0.008)
Educational hypergamy			0.012	(0.010)	−0.016**	(0.005)	0.004	(0.005)	−0.011 +	(0.006)		
Respondent's income share	−0.001***	(0.000)	−0.001***	(0.000)	−0.001***	(0.000)	0.001***	(0.000)	0.001***	(0.000)	0.001***	(0.000)
Partner's standard employment	0.034***	(0.007)	0.046***	(0.008)	0.040***	(0.006)						
Respondent's standard employment							0.033***	(0.006)	0.040***	(0.005)	0.035***	(0.006)
Constant	0.317***	(0.023)	0.394***	(0.027)	0.381***	(0.019)	0.223***	(0.015)	0.259***	(0.013)	0.298***	(0.019)
Observations	2109		1429		2938		2800		4062		2075	
Standard errors in parentheses												

+ $p < 0.10$ * $p < 0.05$ ** $p < 0.01$ *** $p < 0.001$

available to children.[4] Much of the research on diverging destinies focuses on differences in family outcomes by mothers' educational attainment but rarely considers the socioeconomic status of fathers. Changing patterns of educational assortative mating, especially increasing homogamy among the highly educated (Schwartz and Mare 2005), would seem to be directly relevant to this literature, but have received little attention.

Japan is a particularly interesting context to examine potential relationships between educational assortative mating in the parents' generation and children's socioeconomic attainment. Parenting in Japan is characterized by intensive mothering, reflecting both older notions of good mothers (*ryōsai kenbo*) and highly competitive school entrance exams that encourage parental investment in children's extracurricular activities, including cram schools (Brinton 1993; Hirao 2001; Holloway 2010; Tsuya and Choe 2004). The importance of passing these high-stakes exams promotes intensive maternal investment in children's education that may "justify highly educated women's inability to utilize their human capital in the workplace" (Yu 2009: 113). In this context, we might speculate that patterns of educational assortative mating play a limited role in shaping children's outcomes. More specifically, we might expect that the mother's educational attainment is of primary importance regardless of the father's educational level. This is further suggested by abundant empirical evidence of Japanese fathers' limited time spent with children (Ishii-Kuntz et al. 2004; Ishii-Kuntz 2013).

We believe, however, that there are reasons to expect that fathers' education also plays some role in shaping the variation in parents' investment in children's education and thus that there is value in considering patterns of educational pairing as a predictor of children's outcomes. First, there has been a small increase in fathers' time spent with children in recent years (Hertog and Kan 2019). Also, widespread recognition of growing economic inequality (Shirahase 2014; Tachibanaki 2005) may lead parents to increasingly view mothers' and fathers' intensive investment in children's educational success as economically rational (Doepke and Zilibotti 2019). Indeed, a recent study by Hertog and Kan (2019) reported that highly educated fathers have increased their child care time in recent years to a greater degree than less-educated fathers. Furthermore, there is some evidence suggesting that educationally homogamous couples have lower levels of marital conflict and tend to agree on parenting goals that are favorable for children's health and academic achievement (Bai 2018; Beck and González-Sancho 2009; Byun et al. 2020; Pesando 2021; Rauscher 2020). Taken as a whole, these ideas point to the value of examining whether and how educational pairing, net of mothers' and fathers' education level, matters for investments in children's education. We examine the SSM data to address this research question.

[4] Similar ideas have been discussed in other studies (Lundberg et al. 2016; Reeves 2017).

5.4.3.1 Investments in Child's Education

The last question we investigate is the implication of assortative mating for intergenerational mobility. To this end, we examine the relationship between parents' educational pairing and investment in children's education. It is clear from previous research that mothers' educational attainment is positively associated with involvement in their children's education (Holloway et al. 2008; Yamamoto et al. 2006) and children's time spent on studying (Hertog and Zhou 2021). We also know that both fathers' and mothers' educational attainment are associated with their financial investment in children's private education, and with their attitudes toward childrearing (Shirahase 2010). Other research shows that women's labor force participation is negatively associated with children's enrollment in after-school cram schools (*juku*) (Hirao 2007), and women's educational attainment is positively associated with children's participation in preschool (Yamamoto et al. 2006) and afterschool (Tsuya and Choe 2004) programs.

To examine the roles of both parents' educational attainment and their educational pairing pattern for investment in children's education, we used data from SSM 2005 and 2015. Both of these surveys included questions on monthly spending on children's extracurricular activities for parents with a child(ren) in primary or secondary education. We limit our analytical sample to men and women with at least one child aged 6–18. The outcome of interest is spending on extracurricular activities and the main correlates of interest are each parent's educational attainment and their pairing pattern. Similar to what we have done in previous chapters, we classified parents' education into (1) high school or less, (2) junior college/vocational school, and (3) BA + . Reflecting our interest in the potential implications of assortative mating for socioeconomic inequality, we also included a dummy variable that takes the value of 1 if both parents are highly educated (BA +) and zero otherwise. Similar to the JLPS, SSM surveys also target individuals rather than households, and it is likely that the partner primarily responsible for managing household finances (typically wives) will provide more accurate answers to the questions about spending. In this study, therefore, we estimate regression models using a pooled sample of male and female respondents and including a dummy variable to indicate the respondent's gender. Our interest is in the relationship between parents' educational pairing and educational spending on children and we also consider how that relationship may differ depending on parents' employment status and household income. Our focus on parents' employment, especially that of mothers, is motivated by previous studies reporting that married women often engage in paid employment in order to help pay for children's extracurricular activities (Tsuya and Choe 2004).

The relationship between mothers' educational attainment and their employment varies by one's life stage and across time, but recent studies have found a growing positive correlation between women's educational attainment and employment after childbearing (Mugiyama n.d.; Raymo and Iwasawa 2016; Senda 2002). Therefore, a positive association between mothers' educational attainment and spending on children's education may reflect the fact that highly educated women are increasingly more likely to continue their employment, which itself appears to have a negative effect on children's cognitive well-being in Japan mothers' earned income through labor supply is not large enough to compensate for mothers' reduced childcare time.[5] We also include household income in the models in order to examine whether the posited link between higher parental educational attainment and spending on children simply reflects differences in economic resources. Finding a positive correlation between parents' educational attainment and educational investment, net of these structural constraints, would suggest the potential importance of other factors, perhaps including highly educated parents' stronger interest in reproducing their advantaged socioeconomic status by investing in children's education. Descriptive statistics are shown in Table 5.21.

The results presented in Table 5.22 indicate that parents' educational attainment has a non-negligible association with children's educational investment, net of employment status and economic resources. Results show that both mothers' and fathers' educational attainment is positively associated with the amount of spending on children's education, but the size of coefficient for mothers' education is much larger than for fathers' education. For example, Model 1 indicates that husbands' higher education is associated with a 67% increase in educational expenditure as compared with husbands in the lowest educational group, while wife's higher education is associated with a 118% increase in the expenditure, also compared with the lowest educated women. According to Model 2, the coefficient for educational homogamy (both couples have a BA +) is not statistically significant, but the coefficient is negative (–0.66). In this model, the coefficients for mothers' and fathers' higher education (BA +) increased (from 1.18 to 1.64 for mothers and from 0.67 to 0.76 for fathers). As the sum of changes in the coefficients is zero, these results suggest that these three coefficients are offsetting each other, resulting in no positive interaction between mothers' and fathers' educational attainment. Despite that, there is evidence of the additive influence of parents' education on children's educational investment, so we can speculate that numerical increase in educational homogamy especially at the top and bottom of educational distribution may contribute to inequality of educational opportunity. Importantly, these results hold even after adding household annual income (Model 3), suggesting that parents' educational attainment measures something other than the household-level resources available for educational expenditure.

To investigate whether the association between both parents' educational attainment and pairing and educational investment for children changed over time, we also estimated the models separately for SSM 2005 and SSM 2015 focusing on Model 3. The expenditure gap between mothers with a BA + and those with a high school education or less was smaller in the second survey and the coefficient for household income also shrunk across the two surveys. This may reflect the fact that the amount of spending decreased in SSM 2015 as Models 1 to 3 indicate. In contrast, results

---

[5] But see Wang and Raymo (n.d.), which suggests that the negative association between mother's work and children's cognitive ability depends on mother's education. In their analysis, the negative employment effect on child wellbeing is limited to low-educated mothers.

Table 5.21 Descriptive statistics

	Mean	SD	Min	Max
SSM 2015	0.55	0.50	0.00	1.00
Female	0.56	0.50	0.00	1.00
Respondent Age	41.23	5.84	23.00	68.00
Husband HS or less	0.48	0.50	0.00	1.00
Husband JC	0.13	0.34	0.00	1.00
Husband BA +	0.39	0.49	0.00	1.00
Wife HS or less	0.51	0.50	0.00	1.00
Wife JC	0.33	0.47	0.00	1.00
Wife BA +	0.16	0.36	0.00	1.00
Couples with both BA +	0.12	0.33	0.00	1.00
Husband Standard	0.84	0.36	0.00	1.00
Husband Non-standard	0.02	0.15	0.00	1.00
Husband Self-employed	0.13	0.33	0.00	1.00
Husband Not worked	0.01	0.09	0.00	1.00
Wife Standard	0.22	0.42	0.00	1.00
Wife Non-standard	0.38	0.48	0.00	1.00
Wife Self-employed	0.08	0.28	0.00	1.00
Wife Not worked	0.31	0.46	0.00	1.00
Number of children aged from 6 to 18	1.84	0.70	1.00	4.00
Monthly spending on extracurricular activities (log)	22,356.12	21,253.14	0.00	100,000.00
inveslog	8.53	3.40	0.00	11.51
Household annual income	713.91	404.04	30.00	7000.00
Household annual income (log)	12.42	6.36	0.00	18.06
Household annual income missing	0.21	0.41	0.00	1.00
Observations	1807			

show that fathers' higher education, the coefficient for which is statistically insignificant in 2005, is positively and significantly associated with education spending for children in 2015. This may reflect men's increased contribution to parenting in recent years. Similar to the overall results, we did not find a positive association between educational homogamy among the highly educated and expenditures on children's education.

Table 5.22 Regression results for models of the investment in children's extracurricular education (female respondents)

	Model 1		Model 2		Model 3		SSM 2005		SSM 2015	
Respondent Age	0.049***	(0.014)	0.048***	(0.014)	0.044**	(0.014)	0.021	(0.021)	0.062***	(0.019)
Female	−0.082	(0.162)	−0.087	(0.162)	−0.063	(0.161)	−0.256	(0.244)	0.068	(0.217)
Husband JC	0.299	(0.251)	0.279	(0.251)	0.151	(0.250)	0.916 +	(0.495)	0.016	(0.299)
Husband BA +	0.670***	(0.189)	0.759***	(0.202)	0.527*	(0.205)	0.278	(0.297)	0.735*	(0.286)
Wife JC	0.868***	(0.188)	0.843***	(0.189)	0.695***	(0.189)	1.025***	(0.300)	0.452 +	(0.246)
Wife BA +	1.176***	(0.253)	1.639***	(0.441)	1.323**	(0.438)	1.421 +	(0.788)	1.258*	(0.532)
Number of children aged from 6 to 18	0.482***	(0.112)	0.487***	(0.112)	0.438***	(0.111)	0.254	(0.165)	0.591***	(0.152)
SSM 2015	−0.407*	(0.166)	−0.401*	(0.166)	−0.364*	(0.166)				
Couples with both BA +			−0.659	(0.515)	−0.519	(0.509)	−0.347	(0.908)	−0.755	(0.621)
Husband Non-standard					−1.093*	(0.522)	−1.489*	(0.735)	−0.926	(0.744)
Husband Self-employed					−0.341	(0.256)	−0.235	(0.372)	−0.514	(0.355)
Husband Not worked					−1.076	(0.837)	−0.877	(1.008)	−1.641	(1.510)
Wife Non-standard					0.168	(0.213)	0.099	(0.331)	0.252	(0.279)
Wife Self-employed					0.601 +	(0.338)	0.442	(0.492)	0.874 +	(0.472)
Wife Not worked					−0.204	(0.223)	−0.458	(0.335)	0.009	(0.303)
Household annual income (log)					1.066***	(0.193)	1.186***	(0.294)	1.002***	(0.257)
Household annual income missing					16.558***	(3.028)	18.952***	(4.618)	14.955***	(4.042)
Constant	5.134***	(0.623)	5.147***	(0.623)	−11.078***	(3.063)	−11.638*	(4.625)	−11.467**	(4.111)
Observations	1807		1807		1807		807		1000	
R2	0.059		0.060		0.088		0.108		0.093	
Standard errors in parentheses										

+ $p < 0.10$ * $p < 0.05$ ** $p < 0.01$ *** $p < 0.001$

References

Abe, Yukiko, and Akiko S. Oishi. 2007. The role of married women's labor supply on family earnings distribution in Japan. *Journal of Income Distribution* 16 (3–4): 110–127.

Antecol, Heather. 2015. Career and family choices among elite liberal arts graduates. *Demography* 52 (4): 1089–1120. https://doi.org/10.1007/s13524-015-0408-z.

Atkinson, Maxine P., and Becky L. Glass. 1985. Marital age heterogamy and homogamy, 1900 to 1980. *Journal of Marriage and the Family* 47 (3): 685–691.

Bai, Yifan. 2018. The consequences of educational assortative marriage for children's academic proficiency in the United States and China. Unpublished Ph.D. Dissertation, Pennsylvania State University.

van Bavel, Jan. 2012. The reversal of the gender gap in education and female breadwinners in Europe ed. by W. Lutz, K. S. James, V. Skirbekk, and J. van Bavel. *Vienna Yearbook of Population Research* 10: 127–154. https://doi.org/10.1553/populationyearbook2012s127.

van Bavel, Jan, and Martin Klesment. 2017. Educational pairings, motherhood, and women's relative earnings in Europe. *Demography* 54 (6): 2331–2349. https://doi.org/10.1007/s13524-017-0621-z.

van Bavel, Jan, Christine R. Schwartz, and Albert Esteve. 2018. The reversal of the gender gap in education and its consequences for family life. *Annual Review of Sociology* 44 (1): 341–360. https://doi.org/10.1146/annurev-soc-073117-041215.

Van Poppel, Frans, Aart C. Liefbroer, Jeroen K. Vermunt, and Wilma Smeenk. 2001. Love, necessity and opportunity: Changing patterns of marital age homogamy in the Netherlands, 1850–1993. *Population Studies* 55(1):1–13. https://doi.org/10.1080/00324720127681.

Beck, Audrey, and Carlos González-Sancho. 2009. Educational assortative mating and children's school readiness. *Center for Research on Child Wellbeing Working Paper* 1–38.

Becker, Gary S. 1991. *A treatise on the family*. Harvard University Press.

Bonke, Jens, and Gøsta Esping-Andersen. 2011. Family investments in children-productivities, preferences, and parental child care. *European Sociological Review* 27 (1): 43–55. https://doi.org/10.1093/esr/jcp054.

Bozon, Michel. 1991. Women and the age gap between spouses: An accepted domination? *Population* 37.

Breen, Richard, and Signe Andersen. 2012. Educational assortative mating and income inequality in Denmark. *Demography* 49 (3): 867–887. https://doi.org/10.1007/s13524-012-0111-2.

Breen, Richard, and Leire Salazar. 2010. Has increased women's educational attainment led to greater earnings inequality in the United Kingdom? A multivariate decomposition analysis. *European Sociological Review* 26 (2): 143–157. https://doi.org/10.1093/esr/jcp011.

Breen, Richard, and Leire Salazar. 2011. Educational assortative mating and earnings inequality in the United States. *American Journal of Sociology* 117 (3): 808–843. https://doi.org/10.1086/661778.

Brinton, Mary C. 1993. *Women and the economic miracle: Gender and work in postwar Japan*. University of California Press.

Brinton, Mary C., and Sunhwa Lee. 2001. Women's education and the labor market in Japan and South Korea. In *Women's working lives in East Asia*, ed. M. C. Brinton, 125–50. Stanford University Press.

Bueno, Xiana, and Joan García-Román. 2021. Rethinking couples' fertility in Spain: Do partners' relative education, employment, and job stability matter? *European Sociological Review*. https://doi.org/10.1093/esr/jcaa070.

Byun, Soo-yong, Yifan Bai, and Hee Jin Chung. 2020. Consequences of educational assortative marriage for children's academic achievement in South Korea.In *Korean families yesterday and today*, ed. H. Park and H. Woo, 95–118. University of Michigan Press.

Cancian, Maria, and Deborah Reed. 1998. Assessing the effects of wives' earnings on family income inequality. *The Review of Economics and Statistics* 80 (1): 73–79.

Chudnovskaya, Margarita, and Ridhi Kashyap. 2020. Is the end of educational hypergamy the end of status hypergamy? evidence from Sweden. *European Sociological Review* 36 (3): 351–365. https://doi.org/10.1093/esr/jcz065.

Compton, Janice, and Robert A. Pollak. 2007. Why are power couples increasingly concentrated in large metropolitan areas? *Journal of Labor Economics* 25 (3): 475–512. https://doi.org/10.1086/512706.

Costa, Dora L., and Mathew E. Kahn. 2000. Power couples: Changes in the locational choice of the college educated, 1940–1990. *The Quarterly Journal of Economics* 115 (4): 1287–1315. https://doi.org/10.1162/003355300555079.

Doepke, Matthias, and Fabrizio Zilibotti. 2019. *Love, money and parenting: How economics explains the way we raise our kids*. Princeton University Press.

Eika, Lasse, Magne Mogstad, and Basit Zafar. 2019. Educational assortative mating and household income inequality. *Journal of Political Economy* 127 (6): 1–41.

Esping-Andersen, Gøsta. 2007. Sociological explanations of changing income distributions. *American Behavioral Scientist* 50(5), 639–58.

Esteve, Albert, Clara Cortina, and Anna Cabré. 2009. Long term trends in marital age homogamy patterns: Spain, 1922–2006. *Population (English Edition)* 64 (1): 173. https://doi.org/10.3917/pope.901.0173.

Esteve, Albert, Joan Garcia-Roman, and Inaki Permanyer. 2012. The gender-gap reversal in education and its effect on Union formation: The end of hypergamy? *Population and Development Review* 38 (3): 535–546. https://doi.org/10.1111/j.1728-4457.2012.00515.x.

Esteve, Albert, Christine R. Schwartz, Jan van Bavel, Inaki Permanyer, Martin Klesment, and Joan Garcia-Roman. 2016. The end of hypergamy: Global trends and implications. *Population and Development Review* 42 (4): 615–625.

Fujihara, Sho, and Fumiya Uchikoshi. 2019. Declining association with persistent gender asymmetric structure: Patterns and trends in educational assortative marriage in Japan, 1950–1979. *Research in Social Stratification and Mobility* 60: 66–77. https://doi.org/10.1016/j.rssm.2018.12.001.

Fukuda, Setsuya, Shohei Yoda, and Ryohei Mogi. 2021. Educational assortative mating in Japan: Evidence from the 1980–2010 census. *The Journal of Population Studies (Jinkogaku Kenkyu)*. https://doi.org/10.24454/jps.2101001.

Fuwa, Makiko. 2020. Kikon Dansei No Hatarakikata to Shugyo Kankyo (Married Men's Workstyle and Employment Circumstances). In *Deai to Kekkon (Opportunities and Marriage)*, ed. H. Sato and H. Ishida, 103–24. Keiso Shobo. (in Japanese).

Geist, Caludia. 2010. Men's and women's reports about housework. In *Dividing the domestic: Men, women, and household work in cross-national perspective*, ed. J. Treas and S. Drobnič, 217–40. Stanford University Press.

Gornick, Janet C., and Markus Jäntti. 2012. Child poverty in cross-national perspective: Lessons from the luxembourg income study. *Children and Youth Services Review* 34 (3):558–568. https://doi.org/10.1016/j.childyouth.2011.10.016.

Greenwood, Jeremy, Nezih Guner, Georgi Kocharkov, and Cezar Santos. 2014. Marry your like: Assortative mating and income inequality. *The American Economic Review* 104 (5): 1–15.

Greenwood, Jeremy, Nezih Guner, Georgi Kocharkov, and Cezar Santos. 2015. Corrigendum to marry your like: Assortative mating and income inequality. http://Www.Jeremygreenwood.Net/Papers/GgksPandPcorrigendum.Pdf.

Heckert, D. Alex., Thomas C. Nowak, and Kay A. Snyder. 1998. The impact of husbands' and wives' relative earnings on marital disruption. *Journal of Marriage and the Family* 60 (3): 690–703.

Hertog, Ekaterina, and Man Yee Kan. 2019. Educational attainment and domestic work contributions between 1991 and 2016 in Japan. Austin, TX.

Herzberg-Druker, Efrat, and Haya Stier. 2019. "Family matters: The contribution of households' educational and employment composition to income inequality. *Social Science Research* 82: 221–239. https://doi.org/10.1016/j.ssresearch.2019.04.012

Hertog, Ekaterina, and Muzhi Zhou. 2021. Japanese adolescents' time use: The role of household income and parental education. *Demographic Research* 44: 225–238. https://doi.org/10.4054/DemRes.2021.44.9.

Herzberg-, Efrat, and Haya Stier. 2019. Family matters: The contribution of households' educational and employment composition to income inequality. *Social Science Research* 82: 221–239. https://doi.org/10.1016/j.ssresearch.2019.04.012.

Higuchi, Yoshio. 1991. *Nihon Keizai to Shugyo Kodo*. Tokyo: Toyo Keizai Shinposha. (In Japanese).

Higuchi, Yoshio, Kazuyasu Sakamoto, and Risa Hagiwara. 2016. Economic and time constraints on women's marriage, childbirth and employment, and effects of work-life balance policies : Empirical analysis Using Japanese Household Panel Surveys. *Mita Business Review* 58 (6): 29–57 (in Japanese).

Hirao, Keiko. 1999. Josei No Shoki Kyaria Keiseiki Ni Okeru Rodo Shijo Eno Teichakusei (Labor Force Attachment of Women in an Early Career Stage). *The Monthly Journal of the Japan Institute of Labour* 41 (9): 29–41 (in Japanese).

Hirao, Keiko. 2001. Mothers as the best teachers: Japanese motherhood and early childhood education. In *Women's working lives in East Asia*, ed. M. C. Brinton, 180–203. Stanford University Press.

Hirao, Keiko. 2005. Josei No Gakureki to Saishusyoku (Women's Education and labor market Reentry). *Japanese Journal of Family Sociology* 17 (1): 34–43 (in Japanese).

Hirao, Keiko. 2007. The privatized education market and maternal employment in Japan. In *The political economy of Japan's low fertility*, ed. F. M. Rosenbluth, 170–200. Stanford University Press.

Holloway, Susan D. 2010. *Women and family in contemporary Japan*. Cambridge University Press.

Holloway, Susan D., Yoko Yamamoto, Sawako Suzuki, and Jessica D. Mindnich. 2008. Determinants of parental involvement in early schooling: Evidence from Japan. *Early Childhood Research and Practice* 10.

Hornung, Carlton A., and B. Claire McCullough. 1981. Status relationships in dual-employment marriages: Consequences for psychological well-being. *Journal of Marriage and Family* 43 (1): 125–141.

Hryshko, Dmytro, Chinhui Juhn, and Kristin McCue. 2017. Trends in earnings inequality and earnings instability among U.S. couples: How important is assortative matching? *Labour Economics* 48: 168–182. https://doi.org/10.1016/j.labeco.2017.08.006.

Hu, Anning, and Zhenchao Qian. 2015. Educational homogamy and earnings inequality of married couples: Urban China, 1988–2007. *Research in Social Stratification and Mobility* 40: 1–15. https://doi.org/10.1016/j.rssm.2015.01.004.

Ishida, Hiroshi. 2007. Japan: Educational expansion and inequality in access to higher education. In *Stratification in higher education: A comparative study*, ed. Y. Shavit, R. Arum, and A. Gamoran, 63–86. Stanford University Press.

Ishida, Hiroshi. 2013. The transition to adulthood among Japanese Youths: Understanding courtship in Japan. *The ANNALS of the American Academy of Political and Social Science* 646 (1): 86–106. https://doi.org/10.1177/0002716212465589.

Ishida, Hiroshi, and Akira Motegi. 2014. Educational assortative mating among unmarried and married couples in Japan and the United States. *Institute of Social Sciences, the University of Tokyo, Panel Survey Project Discussion Paper Series No. 72.*

Ishii-Kuntz, Masako. 2013. Work environment and Japanese Fathers' involvement in child care. *Journal of Family Issues* 34 (2): 250–269. https://doi.org/10.1177/0192513X12462363.

Ishii-Kuntz, Masako, Katsuko Makino, Kuniko Kato, and Michiko Tsuchiya. 2004. Japanese Fathers of preschoolers and their involvement in child care. *Journal of Marriage and Family* 66 (3): 779–791. https://doi.org/10.1111/j.0022-2445.2004.00052.x.

Kambayashi, Ryo. 2017. *Seiki No Sekai. Hiseiki No Sekai (Standard World, Non-Standard World).* Keio University Press (in Japanese).

Kambayashi, Ryo, Daiji Kawaguchi, and Izumi Yokoyama. 2008. Wage distribution in Japan, 1989–2003. *Canadian Journal of Economics/revue Canadienne D'économique* 41 (4): 1329–1350. https://doi.org/10.1111/j.1540-5982.2008.00506.x.

Kamo, Yoshinori. 2000. 'He said, she said': Assessing discrepancies in husbands' and wives' reports on the division of household labor. *Social Science Research* 29 (4): 459–476. https://doi.org/10.1006/ssre.2000.0674.

Kaneko, Motohisa. 1996. Koto Kyoiku Taishuka No Ninaite (Members of Popularization of Higher Education). *Report on Multimedia Education* 91: 37–59 (in Japanese).

Kawaguchi, Akira. 2002. Douglas-Arisawa Hosoku Ha Yuko Nanoka? (Is Still Douglas-Arisawa's Law Relevant?) *Nihon Rodo Kenkyu Zasshi* 501: 18–21 (in Japanese).

Kawaguchi, Daiji, and Yuko Mori. 2016. Why has wage inequality evolved so differently between Japan and the US? The role of the supply of college-educated workers. *Economics of Education Review* 52: 29–50. https://doi.org/10.1016/j.econedurev.2016.01.002.

Klesment, Martin, and Jan van Bavel. 2017. The reversal of the gender gap in education, motherhood, and women as main earners in Europe. *European Sociological Review* 33 (3): 465–481. https://doi.org/10.1093/esr/jcw063.

Kohara, Miki. 2008. Is the full-time housewife a symbol of a wealthy family? *Japanese Economy* 34 (4): 25–56. https://doi.org/10.2753/JES1097-203X340402.

Kolk, Martin. 2015. Age differences in unions: Continuity and divergence among Swedish couples between 1932 and 2007. *European Journal of Population* 31 (4): 365–382. https://doi.org/10.1007/s10680-015-9339-z.

Kollmeyer, Christopher. 2013. Family structure, female employment, and national income inequality: A cross-national study of 16 Western countries. *European Sociological Review* 29 (4): 816–827. https://doi.org/10.1093/esr/jcs060.

Lundberg, Shelly, Robert A. Pollak, and Jenna Stearns. 2016. Family inequality: Diverging patterns in marriage, cohabitation, and childbearing. *Journal of Economic Perspectives* 30 (2): 79–102. https://doi.org/10.1257/jep.30.2.79.

Mare, Robert D. 1991. Five decades of educational assortative mating. *American Sociological Review* 56 (1): 15–32.

McCall, Leslie, and Christine Percheski. 2010. Income inequality: New trends and research directions. *Annual Review of Sociology* 36 (1): 329–347. https://doi.org/10.1146/annurev.soc.012809.102541.

McLanahan, Sara. 2004. Diverging destinies: How children are faring under the second demographic transition. *Demography* 41 (4): 607–627. https://doi.org/10.1353/dem.2004.0033.

McLanahan, Sara, and Christine Percheski. 2008. Family structure and the reproduction of inequalities. *Annual Review of Sociology* 34 (1): 257–276. https://doi.org/10.1146/annurev.soc.34.040507.134549.

Miller, Rhiannon N. 2020. Educational assortative mating and time use in the home. *Social Science Research* 90:102440. https://doi.org/10.1016/j.ssresearch.2020.102440.

Mincer, Jacob. 1985. Intercountry comparisons of labor force trends and of related developments: An overview. *Journal of Labor Economics* 3 (1): S1–32.

Miwa, Satoshi. 2007. Long-term trends in status homogamy. In *Deciphering stratification and inequality*, ed. Y. Sato, 140–60. Trans Pacific Press.

Monaghan, David. 2015. Income inequality and educational assortative mating: Evidence from the Luxembourg income study. *Social Science Research* 52: 253–269. https://doi.org/10.1016/j.ssresearch.2015.02.001.

Motegi, Akira, and Hiroshi Ishida. 2020. Kekkon Eno Michinori (Pathways into Marriage). In *Deai to Kekkon (Opportunities and Marriage)*, ed. H. Sato and H. Ishida. Keiso Shobo, 44–75 (in Japanese).

Mugiyama, Ryota. n.d. A Reconsideration of Educational Differentials in Women's Employment around Childbirth in Japan.

Mu, Zheng, and Yu Xie. 2014. Marital age homogamy in China: A reversal of trend in the reform era? *Social Science Research* 44: 141–157. https://doi.org/10.1016/j.ssresearch.2013.11.005.

Murakami, Akane. 2001. Kyuju Nendai Ni Okeru Kikon Josei No Shugyo to Shunyu Kakusa (Married Women's Employment and Income Inequality during the 1990s). *Soshioroji* 46 (2): 37–55 (in Japanese).

Nagase, Nobuko. 1999. Work and childbearing choice of married women in Japan: The effect of labor practices. *Journal of Population Problems (Jinko Mondai Kenkyu)* 55 (2): 1–18 (in Japanese).

Nagase, Nobuko. 2004. Standard and nonstandard work arrangements, pay differences, and choice of work by Japanese mothers. In *Nonstandard work in developed economies: Causes and consequences*, ed. S. Houseman and M. Osawa, 267–306. W.E. Upjohn Institute.

National Institute of Population and Social Security Research. 2017. *Marriage and Childbirth in Japan Today: The Fifteenth Japanese National Fertility Survey, 2015*. National Institute of Population and Social Security Research.

Nitsche, Natalie, Anna Matysiak, Jan van Bavel, and Daniele Vignoli. 2018. Partners' educational pairings and fertility across Europe. *Demography* 55(4):1195–1232. https://doi.org/10.1007/s13524-018-0681-8.

Nomes, Eli, and Jan van Bavel. 2016. Marital fertility and educational assortative mating before, during, and after the baby boom in Belgium. *Annales De Démographie Historique* 132 (2): 139–163.

Ohtake, Fumio. 2005. *Nihon No Fubyodo (Inequality in Japan)*. Nihon Keizai Shimbunsha (in Japanese).

Oppenheimer, Valerie Kincade. 1988. A theory of marriage timing. *American Journal of Sociology* 94 (3): 563–591.

Oppenheimer, Valerie Kincade, Matthijs Kalmijn, and Nelson Lim. 1997. Men's career development and marriage timing during a period of rising inequality. *Demography* 34 (3): 311. https://doi.org/10.2307/3038286.

Oshio, Takashi, and Kunio Urakawa. 2014. The association between perceived income inequality and subjective well-being: Evidence from a social survey in Japan. *Social Indicators Research* 116 (3): 755–770. https://doi.org/10.1007/s11205-013-0323-x.

Osiewalska, Beata. 2017. Childlessness and fertility by couples' educational gender (in)Equality in Austria, Bulgaria, and France. *Demographic Research* 37: 325–362. https://doi.org/10.4054/DemRes.2017.37.12.

Pesando, Luca Maria. 2021. A four-country study on the relationship between parental educational homogamy and children's health from infancy to adolescence. *Population Research and Policy Review*. https://doi.org/10.1007/s11113-020-09627-2.

Qian, Zhenchao. 1998. Changes in assortative mating: The impact of age and education, 1970–1990. *Demography* 35 (3): 279–292.

Qian, Yue. 2017. Gender asymmetry in educational and income assortative marriage: Educational and income assortative marriage. *Journal of Marriage and Family* 79 (2): 318–336. https://doi.org/10.1111/jomf.12372.

Qian, Zhenchao, and Daniel T. Lichter. 2018. Marriage markets and intermarriage: Exchange in first marriages and remarriages. *Demography* 55 (3): 849–875. https://doi.org/10.1007/s13524-018-0671-x.

Rauscher, Emily. 2020. Why who marries whom matters: Effects of educational assortative mating on infant health in the United States 1969–1994. *Social Forces* 98 (3): 1143–1173. https://doi.org/10.1093/sf/soz051.

Raymo, James M., and Miho Iwasawa. 2016. *Diverging destinies: The Japanese case*. Springer.

Raymo, James M., and So-Jung Lim. 2011. A new look at married women's labor force transitions in Japan. *Social Science Research* 40 (2): 460–472. https://doi.org/10.1016/j.ssresearch.2010.10.005.

Raymo, James M., and Yu Xie. 2000. Temporal and regional variation in the strength of educational homogamy. *American Sociological Review* 65 (5): 773–781.

Raymo, James M., Miho Iwasawa, and Larry Bumpass. 2009. Cohabitation and family formation in Japan. *Demography* 46 (7): 785–803.

Raymo, James M., Hyunjoon Park, Yu Xie, and Wei-jun Jean Yeung. 2015. Marriage and family in East Asia: Continuity and change. *Annual Review of Sociology* 41 (1): 471–492. https://doi.org/10.1146/annurev-soc-073014-112428.

Reeves, Richard V. 2017. *Dream hoarders: How the American upper middle class is leaving everyone else in the dust, why that is a problem, and what to do about it*. Brookings Institution Press.

Schwartz, Christine R. 2010. Earnings inequality and the changing association between spouses' earnings. *American Journal of Sociology* 115 (5): 1524–1557. https://doi.org/10.1086/651373.

Schwartz, Christine R. 2013. Trends and variation in assortative mating: Causes and consequences. *Annual Review of Sociology* 39 (1): 451–470. https://doi.org/10.1146/annurev-soc-071312-145544.

Schwartz, Christine R., and Hongyun Han. 2014. The reversal of the gender gap in education and trends in marital dissolution. *American Sociological Review* 79 (4): 605–629. https://doi.org/10.1177/0003122414539682.

Schwartz, Christine R., and Robert D. Mare. 2005. Trends in educational assortative marriage from 1940 to 2003. *Demography* 42 (4): 621–646.

Senda, Yukiko. 2002. Influence of childcare resources on the employment continuity of married women: Focusing on occupation and birth cohort. *Journal of Population Problems (Jinko Mondai Kenkyu)* 58 (2): 2–21 (in Japanese).

Shafer, Kevin. 2013. Unique matching patterns in remarriage: Educational assortative mating among divorced men and women. *Journal of Family Issues* 34 (11): 1500–1535. https://doi.org/10.1177/0192513X12459020.

Shirahase, Sawako. 2010. Marriage as an association of social classes in a low fertility rate society. In *Social class in contemporary Japan: Structures, sorting, and strategies*, ed. H. Ishida and D. H. Slater, 57–83. Routledge.

Shirahase, Sawako. 2014. *Social inequality in Japan*. Routledge.

Sweeney, Megan M. 2002. Two decades of family change: The shifting economic foundations of marriage. *American Sociological Review* 67 (1): 132–147. https://doi.org/10.2307/3088937.

Tachibanaki, Toshiaki. 2005. *Confronting income inequality in Japan*. MIT Press.

Tachibanaki, Toshiaki, and Sayaka Sakoda. 2013. *Fufu Kakusa Shakai (Couples Disparity Society)*. Chuo Koronsha (in Japanese).

Taki, Hirofumi. 2019. Kikon Josei No Shugyo Sentaku to Senmon Gakko Gakureki (Married women's labor force Participation and Vocational School Education). Shakai Kagaku Kenkyu 70 (1):31–49 (in Japanese).

Takeuchi, Mamiko. 2004. Josei Shugyo No Paneru Bunseki: Haigusha Shotoku Koka No Saikensho (Panel Analysis of Women's Decision to Work: Re-Examination of the Effects of the Husband's Income). *The Japanese Journal of Labour Studies* 46 (6): 76–88 (in Japanese).

Tanaka, Shigeto. 1997. Kogakurekika to Seibetsu Bungyo (Educational Expansion and Gender Division of Labor). *Japanese Sociological Review* 48 (2): 130–142 (in Japanese).

Tanaka, Yuko, and Yumiko Nishimura. 1986. Shokugyo Keizoku Ni Oyobosu Gakureki Koka [The Effect of Education on Continuous Work]. In *Joshi koto kyoiku no zahyo [Goal of women's education]*, ed. M. Amano. 203–224. Kakiuchi Shuppan. (in Japanese).

Torche, Florencia. 2010. Educational assortative mating and economic inequality: A comparative analysis of three Latin American countries. *Demography* 47 (2): 481–502.

Toyonaga, Kohei. 2021. Kosotsu to Daisotsu No Gakureki Kan Chingin Kakusa Wa Kakudai Shitanoka? (Has Wage Gap Between High School Graduates and University Graduates Increased?). *Kikan Kojin Kinyu (quarterly Personal Finance)* 15 (4): 78–87 (in Japanese).

Treas, Judith. 1987. The effect of women's labor force participation on the distribution of income in the United States. *Annual Review of Sociology* 13: 259–288.

Tsutsui, Junya. 2016. *Kekkon to Kazoku No Korekawa (The Future of Marriage and Family)*. Kobunsha. (in Japanese).

Tsuya, Noriko O, and Minja Kim, Choe. 2004. Investments in children's education, desired fertility, and women's employment. In *Marriage, work, and family in comparative perspective*, ed. N. O. Tsuya and L. L. Bumpass, 76–94.

Tsuya, Noriko O., Larry L. Bumpass, and Minja Choe. 2000. Gender, employment, and housework in Japan, South Korea, and the United States. *Review of Population and Social Policy* 9: 195–220.

Tsuya, Noriko O., Larry L. Bumpass, Minja Choe, and Ronald R. Rindfuss. 2012. Employment and household tasks of Japanese couples, 1994–2009. *Demographic Research* 27: 705–718. https://doi.org/10.4054/DemRes.2012.27.24.

Tzeng, Meei-Shenn. 1992. The effects of socioeconomic heterogamy and changes on marital dissolution for first marriages. *Journal of Marriage and Family* 54 (3): 609–619.

Uchikoshi, Fumiya. 2018. A consequence of assortative mating through changes in a couple's household income: Focusing on a couple's educational pairing and the wife's labor market participation. *Japanese Journal of Family Sociology* 30 (1): 18–30 (in Japanese).

Uchikoshi, Fumiya. 2019. Consequences of educational assortative mating through divorce. *Japanese Sociological Review* 70 (1): 10–26 (in Japanese). https://doi.org/10.4057/jsr.70.10.

Uchikoshi, Fumiya. 2020. Explaining declining educational homogamy: The role of institutional changes in higher education in Japan. *SocArxiv* 45.https://doi.org/10.31235/osf.io/45axy

Wakisaka, Akira, and Tomita Yasunobu. 2001. *Daisotsu Josei No Hataraki Kata (Workstyle of College Educated Women)*. Nihon Rodo Kenkyu Kiko. (in Japanese).

Waldfogel, Jane, Yoshio Higuchi, and Masahiro Abe. 1999. Family leave policies and women's retention after childbirth: Evidence from the United States, Britain, and Japan. *Journal of Population Economics* 12 (4): 523–545. https://doi.org/10.1007/s001480050112

Wang, Jia, and James M. Raymo. n.d. The long-term effect of maternal employment on child well-being in Japan.

West, Candace, and Don H. Zimmermann. 1987. Doing gender. *Gender and Society* 12:125–51.

Western, Bruce, Deirdre Bloome, and Christine Percheski. 2008. Inequality among American families with children, 1975 to 2005. *American Sociological Review* 73 (6): 903–920. https://doi.org/10.1177/000312240807300602.

Yamada, Ken, and Daiji Kawaguchi. 2015. The changing and unchanged nature of inequality and seniority in Japan. *The Journal of Economic Inequality* 13 (1): 129–153. https://doi.org/10.1007/s10888-014-9295-6.

Yamamoto, Yoko, Susan D. Holloway, and Sawako Suzuki. 2006. Maternal involvement in preschool children's education in Japan: Relation to parenting beliefs and socioeconomic status. *Early Childhood Research Quarterly* 21 (3): 332–346. https://doi.org/10.1016/j.ecresq.2006.07.008.

Yonezawa, Akiyoshi. 2008. The expansion and differentiation of the Japanese higher education system and educational attainment. In *The 2005 SSM research series volume 5: structure of educational attainment*, ed. A. Yonezawa, 113–39. 2005 SSM Survey Committee.

Yu, Wei-hsin. 2009. *Gendered trajectories*. Stanford University Press.

Zagel, Hannah, and Richard Breen. 2019. Family demography and income inequality in West Germany and the United States. *Acta Sociologica* 62 (2): 174–192. https://doi.org/10.1177/0001699318759404.

Chapter 6
Conclusion and Future Directions

In this chapter, we summarize and evaluate the results of our empirical analyses. We discuss these results with reference to previous studies on Japan as well as from a comparative perspective. We also touch on further questions that we believe should be addressed in future research.

6.1 What We Know (and Don't Know) from Research on Japan

6.1.1 What We Know

Our main goal in the empirical analyses was to document trends and patterns of assortative mating in Japan, and to examine their potential consequences. Many of our findings were consistent with some earlier studies, but not consistent with expectations based on theory and/or other past research. First, we examined the assortative mating process with a focus on evaluating the winnowing hypothesis. Results were consistent with our theoretical expectations. Compared with dating couples, married couples tend to be more educationally homogamous. This is similar to the findings of earlier work by Motegi and Ishida (2020), but only for highly educated groups. At the lower end of the educational distribution, we found that marriages are less likely to be homogamous than are dating relationships. Also consistent with expectations from the winnowing hypothesis, we found that relationship duration is positively associated with the likelihood of educational homogamy. From this result, we can speculate that increasingly long periods of dating in contemporary Japan may contribute to more educationally homogamous marriages, net of other forces that may work in the same or opposite direction. For example, later ages at marriage that accompany long periods of dating mean a longer duration from school graduation, a factor that is theoretically linked to less homogamy (Mare 1991). Future research should thus

F. Uchikoshi and J. M. Raymo, *Educational Assortative Mating in Japan*,
Population Studies of Japan, https://doi.org/10.1007/978-981-16-3713-1_6

carefully distinguish these potentially countervailing influences on the prevalence of educational homogamy (or patterns of educational pairing, more generally).

Second, we examined trends in educational assortative mating. Consistent with a large number of previous studies (Fujihara and Uchikoshi 2019; Fukuda et al. 2021; Miwa 2007; Raymo and Xie 2000), our results indicate that the prevalence of educational homogamy has increased while the odds of educational homogamy, i.e., relative rates net of compositional change, have declined. Meanwhile, educational hypogamy has increased, in both absolute and relative terms. As we discuss below, the declining odds of educational homogamy in Japan contrasts with evidence of increasing educational homogamy in several other countries, suggesting the need to consider cross-national differences in the complex array of social and economic forces that influences patterns of educational assortative mating.

Third, we examined two related questions regarding the implications of assortative mating for income inequality. Somewhat surprisingly, and in contrast to the results of prior studies (Mugiyama n.d.; Raymo and Iwasawa 2016; Senda 2002), we found no evidence that the educational gradient in married women's employment after childbearing has increased in recent years. We speculated that this inconsistency may reflect differences across surveys in the measurement of employment, especially the classification of women who take a short break/leave from their job around childbirth. Our results also show, again somewhat surprisingly, that husbands' education is not significantly associated with married women's employment. This finding could simply reflect our inability to adequately control for economic resources at the household level (e.g., household income), but it could also have potentially important implications for our understanding of inequality to the extent that educational homogamy among the highly educated is no longer associated with reductions in the incentive for married women to continue their employment.

However, our counterfactual analyses indicate that changing spouse pairing patterns over the past three decades have mitigated the growth in income inequality; inequality would be even greater if there had been no increase in educational homogamy or in married women's employment. Of particular importance is the increasing prevalence of educationally hypogamous couples, which are characterized by relatively high employment rates for mothers, whose earnings work to equalize the income distribution by augmenting the relatively low earnings of their husbands. Among educationally homogamous couples at high levels of educational attainment, the disequalizing increase in career-oriented women in "power couples" (Compton and Pollak 2007; Costa and Kahn 2000; Tachibanaki and Sakoda 2013) is offset by a relatively high prevalence of couples in which wives interrupt their employment to focus on childrearing. In contrast, educational hypergamy, which has declined in prevalence, is characterized by a higher proportion of non-working women, contributing to a larger income variation within this group and hence to overall inequality. To summarize, despite an increase in highly educated homogamous couples and educationally hypogamous couples, there is little evidence that these changes have contributed to growing income inequality in Japan. It is, of course, possible that the rise in highly educated homogamous "power couples" could contribute to increased income inequality, but our results indicate that the majority of highly educated homogamous couples are not two high earners, but something closer to one and a half earner couples, with the wife focusing on childrearing while contributing some supplementary income.

Our results regarding family outcomes are largely inconsistent with the findings of earlier research. First, in contrast to the expectation that educationally hypogamous couples may have fewer children to accommodate wives' role as an equal earner (Nomes and van Bavel 2016), we found that educational homogamy is associated with fewer children among highly educated women. One possible explanation for this finding is that highly educated homogamous couples may be selective of parents who prioritize the quality of children over the quantity (Becker 1991). Also, in contrast to previous findings on the declining relative risk of divorce among educationally hypogamous couples (Schwartz and Han 2014; Uchikoshi 2019), our results indicate that these couples are more likely to experience divorce than other educational pairings. This new finding reflects our decision to limit our analytical focus to those who can actually marry hypogamously, namely middle- or highly educated women. Also, we did not find evidence to support the equalization hypothesis that husbands will perform a higher share of housework in hypogamous marriages. Finally, we did not find a positive association between educational assortative mating and expenditures on children's education.

6.1.2 *What We Don't Know*

Because some of our findings are consistent with prior studies while others are not, it is important to consider how to contextualize our results in a comparative perspective. The clear trend toward less educational homogamy, net of compositional change, contrasts with trends in countries like the United States, where an increase in economic sorting has led to more educational homogamy at the top of the educational distribution (Schwartz and Mare 2005). An increase in educational homogamy has been observed primarily in countries where women's access to both education and the labor market has improved in recent decades (Han 2010; Ravazzini et al. 2017; Wong 2003), but trends similar to those observed in Japan have been found in France (Bouchet-Valat 2014), Denmark (Andrade and Thomsen 2019), and in Eastern European (Katrňák and Manea 2020) and East Asian (Smits and Park 2009) countries. Future studies thus need to carefully consider the social and economic forces underlying cross-national variation in educational assortative mating trends.

Another contribution to the comparative literature on assortative mating is our evaluation of the impacts of educational pairing on income inequality. We argued that it is critical to consider how women's labor market context may shape the relationship between educational assortative mating and income inequality. While previous research suggests that the implications of educational assortative mating for income inequality vary depending on the strength of the association between women's educational attainment and employment, especially full-time employment

(Herzberg-Druker and Stier 2019; Schwartz 2013), our evidence from Japan points to the importance of recognizing that the role of women's employment also depends on the type of work women doafter childbearing. An implicit assumption in previous studies is that an increase of highly educated women who marry homogamously and continue full-time employment may contribute to growing income inequality. While this makes good sense in societies characterized by rising rates of married women's labor force participation such as Denmark (Breen and Andersen 2012; Esping-Andersen 2007), married women's employment in Japan is often motivated by the need to support the male breadwinner's earnings and thus works to reduce income inequality.

Lastly, our analyses of the relationships between educational assortative mating and several individual-level outcomes, including family formation and intergenerational mobility, also provide little evidence of a strong role for educational assortative mating. With a few exceptions in our analyses of fertility and divorce, our findings suggest that individual outcomes are largely unrelated to educational pairing patterns. These results stand in contrast to those from other countries (Bai 2018; Beck and González-Sancho 2009; Byun et al. 2020; Miller 2020; Pesando 2021; Rauscher 2020), again suggesting that linkages between educational assortative mating and life outcomes vary across institutional contexts. Because research on these topics remains limited, we see tremendous value in careful cross-national research on the various ways in which educational pairing patterns may be associated with life outcomes.

6.2 Future Directions for Comparative Research

Research on growing access to higher education, the increasing value of educational attainment in the labor market, and the reversal of the gender gap in higher education increasingly emphasizes trends, causes, and consequences of educational assortative mating, but many important questions remain unanswered. In this section, we review these questions, focusing particularly on those that concern understanding population heterogeneity and those that concern detecting causal relationships and their underlying mechanisms.

6.2.1 Understanding Population Heterogeneity

Demographers are keenly interested in how heterogeneity within the population affects demographic outcomes (Xie 2000). It is thus interesting that studies on educational assortative mating tend to focus on a relatively homogeneous population, i.e., first-married couples. The increasing prevalence of higher-order marriages makes it critical to ask whether patterns of assortative mating differ by the order of marriage, and if so, how do they differ and what are the factors that underlie those differences (Choi and Tienda 2018; Hu and Qian 2019; Shafer 2013; Shafer and James 2013).

Immigration is another important source of growing population heterogeneity and it is clear that patterns of educational assortative mating among immigrants differ from those of the native-born population (Choi and Mare 2012; Choi and Tienda 2017; Qian and Lichter 2011; van Zantvliet et al. 2014). Importantly, citizenship and residential status are potentially valuable sources of insight into how educational attainment intersects with other statuses in the creation of social boundaries in the marriage market (Qian and Qian 2017). One methodological difficulty in this research is the need to construct measures of education attained by immigrants in their countries of origin that are comparable to conventional categories of educational attainment in their destination countries. Because prior work on intergenerational educational mobility of immigrants has argued that these measures of educational attainment are often not comparable (Feliciano and Lanuza 2017), future studies need to think hard about how we can incorporate growing populations from different educational contexts into assortative mating studies.

6.2.2 Mechanisms and Consequences

In general, studies on assortative mating have tended to focus on associational rather than causal mechanisms (see Musick et al. 2012 and Qian et al. 2005 for exceptions). Similarly, studies that examine the impacts of educational expansion on educational assortative mating using experimental designs are relatively limited (Andersson 2019; Rauscher 2015). It is clear that stronger causal evidence is needed to rigorously test expectations derived from modernization theory or other widely referenced frameworks (e.g., structural theory). This point was clearly articulated in a recent review of research on the reversal in the gender gap in educational attainment in which the authors stressed the value of understanding whether the reversal in the gender gap has causally contributed to the observed rise in educational hypogamy (van Bavel et al. 2018). The question of causal mechanisms is further complicated by the growing selectivity of marriage. We stress the importance of addressing these issues so that we can evaluate the theoretical relevance and empirical implications of investigating the role of educational assortative mating in the creation of inequality and social change.

6.2.2.1 Growing Institutional Heterogeneity Within Higher Education

Past research has demonstrated that trends in educational assortative mating, net of compositional change, depend on institutional context (Blossfeld 2009; Blossfeld and Timm 2003; Hout and DiPrete 2006; Rauscher 2015). Among the possible institutional mechanisms that might account for mixed findings on trends in educational assortative mating, educational expansion is likely of particular importance. However, focusing simply on levels of higher educational attainment is limiting in that it does not consider how increasing educational attainment may be accompanied by concurrent growth in institutional differentiation. Of particular importance is growth in institutional heterogeneity through the proliferation of lower-tier institutions of higher education (Arum et al. 2007; Hoxby 2009). This is an important omission in the assortative mating literature given the accumulation of research on the effects of horizontal stratification (i.e., heterogeneity within the same years of education) on a range of outcomes (Gerber and Cheung 2008).

The basic concern is that an implicit assumption of a uniform increase in access to higher education is violated when the expansion of higher education is accompanied by growing variation in its nature and quality (Arum et al. 2007; Ayalon et al. 2008; Hoxby 2009). Previous research on stratification in higher education has argued that this differentiation is often driven by an increase in lower-tier institutions (Arum et al. 2007: 5). Ignoring potential growth of within-group variation in higher education is clearly problematic for research on educational assortative mating and more attention to this issue may help us to better understand the mixed evidence in recent research on the trends in educational homogamy.

Analyses of the role of college selectivity in the marriage market in the United States concluded that college graduates tend to marry spouses from post-secondary institutions with similar academic selectivity (Arum et al. 2008; Ford 2020) and that graduates of prestigious schools are more likely than graduates of less prestigious schools to marry spouses with a college degree (Arum and Roksa 2014). These findings suggest that graduates of prestigious institutions are more likely to seek college graduates of similarly ranked institutions to maintain social boundaries. Although previous studies mentioned this pattern as a likely explanation for recent trends in educational homogamy (Arum et al. 2008; Schwartz and Mare 2005), they did not examine this possibility explicitly by modeling the implications of the growing differentiation in higher education for trends in educational assortative mating.

Attention to the growing heterogeneity among university graduates in terms of institutional selectivity may play a key role in helping us to understand mixed evidence on trends in educational assortative mating across countries. This may be particularly important in the Japanese context, where increased post-secondary educational attainment has been promoted through the growth in private institutions typically located low on the hierarchy of selectivity (Fujihara and Ishida 2016; Ishida 2007). The resulting growth in the heterogeneity of university graduates with respect to institutional selectively and prestige may mean that graduates from lower-tier institutions are more likely to marry spouses without a university degree, while preferences for educational homogamy among prestigious university graduates remain strong or even increase.[1] This is supported by our results (above) using data on spouse pairing patterns with respect to college selectivity to demonstrate substantial differences in the odds of homogamy by institutional prestige.

[1] A few recent studies have begun to grapple with this question, interpreting evidence of overall growth in educational hypogamy, combined with an increase in educational homogamy among highly educated women, as evidence that "the social and economic boundaries between lower-ranked universities and technical colleges may be declining among younger cohorts" (Fukuda et al. 2020: 1393).

Institutional differentiation accompanying higher education expansion has also been documented in other countries (Uchikoshi 2020). Future research should thus take a cross-national comparative approach to examining whether the growing institutional heterogeneity in higher education is associated with changes in educational assortative mating. Analyzing how spouse pairing patterns are influenced by educational expansion and differentiation would provide new insights into our understanding of the role of educational attainment in processes of social stratification. We hope that our preliminary analyses on this point in Chap. 5 will motivate further research on this issue.

6.2.2.2 Growing Selectivity of Marriage

The rapid increase in the never-married population in Japan implies that the married population is increasingly selective, an issue of critical importance for research on trends in educational assortative mating. What does the increasing selectivity of marriage mean for our ability to compare patterns of assortative mating across time (Fukuda et al. 2019; Yoshida 2011)? We see value in carefully considering this question, especially in light of the observed decline in educational homogamy in Japan, the opposite of what we see in some other countries including the United States. For example, it is possible that the rapid increase in less-educated men who have never married (Fukuda et al. 2020) reflects the growing difficulty of these low-educated men to marry (homogamously). It would also be interesting to examine the potential consequences of the increase in never-married singles, especially low-SES men, for other stratification or demographic outcomes, including income inequality, fertility or mortality trends, or social security. While in some sections we paid attention to the population at risk of marriage, this study mostly focused on the married population, which precludes investigation of these questions. We thus advocate for research that integrates information on both marriage timing (risk) and patterns of assortative mating.

6.2.2.3 Evaluating the Consequences of Educational Assortative Mating

In sum, there is clearly an abundance of evidence suggesting that macro-level social changes, including educational expansion (Rauscher 2015), reversal in the gender gap in education (Esteve et al. 2016; van Bavel et al. 2018), and the "gender revolution" (Goldscheider et al. 2015), have important implications for patterns and trends in educational assortative mating. In contrast, our understanding of how educational assortative mating shapes society is quite limited. On the one hand, it is true that previous studies have provided empirical evidence regarding (changing) relationships between educational assortative mating and a wide range of outcomes, including income inequality (Schwartz 2010), intergenerational mobility (Mare 2016), and demographic outcomes such as divorce (Schwartz and Han 2014) and fertility

(Nomes and van Bavel 2016). On the other hand, however, it is still not clear how these outcomes interact with each other in ways that shed light on how changing patterns of assortative mating contribute to social change. For example, it is certainly plausible that changing gender norms and expectations of more egalitarian spousal relationships (Goldscheider et al. 2015) contribute to changes in incentives and bases for marriage formation (Oppenheimer 1997) that may result in more positive assortative mating by socioeconomic resources, with attendant implications for rising income inequality (Breen and Andersen 2012) or health inequality (Rauscher 2020).

This scenario accords with McLanahan's (2004)"diverging destinies" thesis, which posits a growing gap in children's life course outcomes driven by socioeconomic bifurcation in parents' family behavior. Although assortative mating has not typically been an explicit focus in research on diverging destinies, it is certainly plausible that an increase in educational homogamy, driven by the rising economic potential of highly educated women, may exacerbate the polarization of children's resources and opportunities. While some previous studies provide convincing evidence of this, others suggest that the relationship between educational assortative mating and inequality is not so straightforward (Breen and Salazar 2011; Goñalons-Pons and Schwartz 2017). Also, it needs to be recognized that we still lack a comprehensive theoretical framework and associated empirical evidence with which to understand how assortative mating contributes to inequality and stratification. Further attention to the processes through which spouse pairing patterns shape opportunities for children's health (Pesando 2021; Rauscher 2020), material and emotional resources, educational success, and adult outcomes is clearly needed.

6.2.3 Concluding Remarks

Our efforts to synthesize the potential impacts of educational assortative mating provide some limited support for changing spouse pairing patterns as an explanation for growing inequality and family change. However, we also argued that there are modest signs of change, suggesting that the role of spouse pairing patterns for stratification and social change in Japan is likely to be of greater importance in the coming years. A number of studies point to changes in spouse selection in the marriage market (Brinton et al. 2021; Fukuda 2013; Fukuda et al. 2020; IPSS 2017), possibly reflecting growing economic precarity among men (Brinton 2011), women's improved access to higher education and the labor market (Raymo and Iwasawa 2016), and rising income inequality (Shirahase 2014; Tachibanaki 2005). Gender norms also appear to be changing, with greater support for more egalitarian relationships (Choe et al. 2014), although the speed of change is slow (Brinton and Oh 2019).

These changing social and family contexts favor dual-earner households in which both spouses contribute to market and household work. The implications of such change may be mixed, with the rise of "power couples" contributing to growing inequality and less social mobility (Raymo and Iwasawa 2016; Tachibanaki and

Sakoda 2013; Tsutsui 2016), while also resulting in a more equal gender division of labor within couples (Fukuda et al. 2020). These changes may also lead to further increase in educational hypogamy (Fukuda et al. 2021) and reduced social stigma toward couples in which the wife is more highly educated or outearns her husband (Uchikoshi 2019). For all of these reasons, future research on assortative mating should pay close attention to ongoing demographic changes with respect to educational attainment, female employment, and family formation.

References

Andersson, Linus. 2019. Do the formative aspects of education really matter for educational assortative mating? Cues from a natural experiment. *Research in Social Stratification and Mobility* 64:100435. https://doi.org/10.1016/j.rssm.2019.100435.

Andrade, Stefan B., and Jens-Peter. Thomsen. 2019. Educational assortative mating: A micro-educational approach. *The British Journal of Sociology* 70 (4): 1245–1275. https://doi.org/10.1111/1468-4446.12488.

Arum, Richard, and Josipa Roksa. 2014. *Aspiring adults adrift: Tentative transitions of college graduates*. University of Chicago Press.

Arum, Richard, Adam Gamoran, and Yossi Shavit. 2007. More inclusion than diversion: Expansion, differentiation, and market structure in higher education. In *Stratification in higher education: A comparative study*, ed. Y. Shavit, R. Arum, and A. Gamoran, 1–35. Stanford University Press.

Arum, Richard, Josipa Roksa, and Michelle J. Budig. 2008. The romance of college attendance: Higher education stratification and mate selection. *Research in Social Stratification and Mobility* 26 (2): 107–121. https://doi.org/10.1016/j.rssm.2008.02.003.

Ayalon, Hanna, Eric Grodsky, Adam Gamoran, and Abraham Yogev. 2008. Diversification and inequality in higher education: A comparison of Israel and the United States. *Sociology of Education* 81 (3): 211–241. https://doi.org/10.1177/003804070808100301.

Bai, Yifan. 2018. The consequences of educational assortative marriage for children's academic proficiency in the United States and China. Unpublished Ph.D. Dissertation, Pennsylvania State University.

Beck, Audrey, and Carlos González-Sancho. 2009. Educational assortative mating and Children's School readiness. *Center for Research on Child Wellbeing Working Paper* 1–38.

Becker, Gary S. 1991. *A treatise on the family*. Harvard University Press.

Blossfeld, Hans-Peter. 2009. Educational assortative marriage in comparative perspective. *Annual Review of Sociology* 35 (1): 513–530. https://doi.org/10.1146/annurev-soc-070308-115913.

Blossfeld, Hans-Peter., and Andreas Timm. 2003. *Who marries whom?*. Springer.

Bouchet-Valat, Milan. 2014. Changes in educational, social class and social class of origin homogamy in France (1969–2011): Greater openness overall but increased closure of elites. *Revue Française De Sociologie* 55 (3): 324–364.

Breen, Richard, and Signe Andersen. 2012. Educational assortative mating and income inequality in Denmark. *Demography* 49 (3): 867–887. https://doi.org/10.1007/s13524-012-0111-2.

Breen, Richard, and Leire Salazar. 2011. Educational assortative mating and earnings inequality in the United States. *American Journal of Sociology* 117 (3): 808–843. https://doi.org/10.1086/661778.

Brinton, Mary C. 2011. *Lost in transition: Youth, work, and instability in postindustrial Japan*. Cambridge University Press.

Brinton, Mary C., and Eunsil Oh. 2019. Babies, work, or both? Highly educated women's employment and fertility in East Asia. *American Journal of Sociology* 125 (1): 105–140. https://doi.org/10.1086/704369.

Brinton, Mary C., Eunmi Mun, and Ekaterina Hertog. 2021. Singlehood in contemporary Japan: Rating, dating, and waiting for a good match. *Demographic Research* 44: 239–276. https://doi.org/10.4054/DemRes.2021.44.10.

Byun, Soo-yong, Yifan Bai, and Hee Jin Chung. 2020. Consequences of educational assortative marriage for children's academic achievement in South Korea. In *Korean families yesterday and today*, ed. H. Park and H. Woo, 95–118. University of Michigan Press.

Choi, Kate H., and Robert D. Mare. 2012. International migration and educational assortative mating in Mexico and the United States. *Demography* 49 (2): 449–476. https://doi.org/10.1007/s13524-012-0095-y.

Choe, Minja Kim, Larry L. Bumpass, Noriko O. Tsuya, and Ronald R. Rindfuss. 2014. Nontraditional family-related attitudes in Japan: Macro and Micro determinants. *Population and Development Review* 40 (2): 241–71. https://doi.org/10.1111/j.1728-4457.2014.00672.x.

Choi, Kate H., and Marta Tienda. 2017. Marriage-market constraints and mate-selection behavior: Racial, ethnic, and gender differences in intermarriage: Marriage market and intermarriage. *Journal of Marriage and Family* 79 (2): 301–317. https://doi.org/10.1111/jomf.12346.

Choi, Kate H., and Marta Tienda. 2018. Intermarriage and the lifecycle timing of migration. *International Migration Review* imre.12326. https://doi.org/10.1111/imre.12326.

Compton, Janice, and Robert A. Pollak. 2007. Why are power couples increasingly concentrated in large metropolitan areas? *Journal of Labor Economics* 25 (3): 475–512. https://doi.org/10.1086/512706.

Costa, Dora L., and Mathew E. Kahn. 2000. Power couples: Changes in the locational choice of the college educated, 1940–1990. *The Quarterly Journal of Economics* 115 (4): 1287–1315. https://doi.org/10.1162/003355300555079.

Esping-Andersen, Gøsta. 2007. Sociological explanations of changing income distributions. *American Behavioral Scientist* 50 (5): 639–658.

Esteve, Albert, Christine R. Schwartz, Jan van Bavel, Inaki Permanyer, Martin Klesment, and Joan Garcia-Roman. 2016. The end of hypergamy: Global trends and implications. *Population and Development Review* 42(4), 615–25.

Feliciano, Cynthia, and Yader Lanuza. 2017. An Immigrant paradox? Contextual attainment and intergenerational educational mobility. *American Sociological Review* 82 (1): 211–241. https://doi.org/10.1177/0003122416684777.

Ford, Karly Sarita. 2020. Marrying within the alma mater: Understanding the role of same-university marriages in educational homogamy. *Sociological Research Online* 25 (2): 254–272. https://doi.org/10.1177/1360780419867710.

Fujihara, Sho, and Hiroshi Ishida. 2016. The absolute and relative values of education and the inequality of educational opportunity: Trends in access to education in postwar Japan. *Research in Social Stratification and Mobility* 43: 25–37. https://doi.org/10.1016/j.rssm.2016.03.001.

Fujihara, Sho, and Fumiya Uchikoshi. 2019. Declining association with persistent gender asymmetric structure: Patterns and trends in educational assortative marriage in Japan, 1950–1979. *Research in Social Stratification and Mobility* 60: 66–77. https://doi.org/10.1016/j.rssm.2018.12.001.

Fukuda, Setsuya. 2013. The changing role of women's earnings in marriage formation in Japan. *The ANNALS of the American Academy of Political and Social Science* 646 (1): 107–128. https://doi.org/10.1177/0002716212464472.

Fukuda, Setsuya, Shohei Yoda, and Ryohei Mogi. 2021. Educational assortative mating in Japan: Evidence from the 1980–2010 census. *The Journal of Population Studies (Jinkogaku Kenkyu)*. https://doi.org/10.24454/jps.2101001.

Fukuda, Setsuya, James M. Raymo, and Shohei Yoda. 2020. Revisiting the educational gradient in marriage in Japan. *Journal of Marriage and Family* 82 (4): 1378–1396. https://doi.org/10.1111/jomf.12648.

Gerber, Theodore P., and Sin Yi Cheung. 2008. Horizontal stratification in postsecondary education: Forms, explanations, and implications. *Annual Review of Sociology* 34 (1): 299–318. https://doi.org/10.1146/annurev.soc.34.040507.134604.

Goldscheider, Frances, Eva Bernhardt, and Trude Lappegård. 2015. The gender revolution: A framework for understanding changing family and demographic behavior. *Population and Development Review* 41 (2): 207–239. https://doi.org/10.1111/j.1728-4457.2015.00045.x.

Goñalons-Pons, Pilar, and Christine R. Schwartz. 2017. Trends in economic homogamy: Changes in assortative mating or the division of labor in marriage? *Demography* 54 (3): 985–1005. https://doi.org/10.1007/s13524-017-0576-0.

Han, Hongyun. 2010. Trends in educational assortative marriage in China from 1970 to 2000. *Demographic Research* 22: 733–770. https://doi.org/10.4054/DemRes.2010.22.24.

Herzberg-, Efrat, and Haya Stier. 2019. Family matters: The contribution of households' educational and employment composition to income inequality. *Social Science Research* 82: 221–239. https://doi.org/10.1016/j.ssresearch.2019.04.012.

Hout, Michael, and Thomas A. DiPrete. 2006. What we have learned: RC28's contributions to knowledge about social stratification. *Research in Social Stratification and Mobility* 24 (1): 1–20. https://doi.org/10.1016/j.rssm.2005.10.001.

Hoxby, Caroline M. 2009. The changing selectivity of American colleges. *Journal of Economic Perspectives* 23 (4): 95–118.

Hu, Yang, and Yue Qian. 2019. Educational and age assortative mating in China: The importance of marriage order. *Demographic Research* 41: 53–82. https://doi.org/10.4054/DemRes.2019.41.3.

Ishida, Hiroshi. 2007. Japan: Educational expansion and inequality in access to higher education. In *Stratification in higher education: A comparative study*, ed. by Y. Shavit, R. Arum, and A. Gamoran, 63–86. Stanford University Press.

Katrňák, Tomáš, and Beatrice Chromková Manea. 2020. Change in prevalence or preference? Trends in educational homogamy in six european countries in a time of educational expansion. *Social Science Research* (Online first).

Mare, Robert D. 1991. Five decades of educational assortative mating. *American Sociological Review* 56 (1): 15–32.

Mare, Robert D. 2016. Educational homogamy in two gilded ages: Evidence from inter-generational social mobility data. *The ANNALS of the American Academy of Political and Social Science* 663 (1): 117–139. https://doi.org/10.1177/0002716215596967.

McLanahan, Sara. 2004. Diverging destinies: How children are faring under the second demographic transition. *Demography* 41 (4): 607–627. https://doi.org/10.1353/dem.2004.0033.

Miller, Rhiannon N. 2020. Educational assortative mating and time use in the home. *Social Science Research* 90:102440. https://doi.org/10.1016/j.ssresearch.2020.102440.

Miwa, Satoshi. 2007. Long-term trends in status homogamy. In *Deciphering stratification and inequality*, edited by Y. Sato, 140–60. Trans Pacific Press.

Motegi, Akira, and Hiroshi Ishida. 2020. Kekkon Eno Michinori (Pathways into Marriage). In *Deai to Kekkon (Opportunities and Marriage)*, edited by H. Sato and H. Ishida, 44–75. Keiso Shobo.

Musick, Kelly, Jennie E. Brand, and Dwight Davis. 2012. Variation in the relationship between education and marriage: Marriage market mismatch? *Journal of Marriage and Family* 74 (1): 53–69. https://doi.org/10.1111/j.1741-3737.2011.00879.x.

National Institute of Population and Social Security Research. 2017. *Marriage and Childbirth in Japan Today: The Fifteenth Japanese National Fertility Survey, 2015*. National Institute of Population and Social Security Research.

Nomes, Eli, and Jan van Bavel. 2016. Marital fertility and educational assortative mating before, during, and after the baby boom in Belgium. *Annales De Démographie Historique* 132 (2): 139–163.

Oppenheimer, Valerie Kincade. 1997. Women's employment and the gain to marriage: The specialization and trading model. *Annual Review of Sociology* 23: 431–453.

Pesando, Luca Maria. 2021. A four-country study on the relationship between parental educational homogamy and children's health from infancy to adolescence. *Population Research and Policy Review*. https://doi.org/10.1007/s11113-020-09627-2.

Qian, Zhenchao, and Daniel T. Lichter. 2011. Changing patterns of interracial marriage in a multiracial society. *Journal of Marriage and the Family* 73 (5): 1065–1084. https://doi.org/10.1111/j.1741-3737.2011.00866.x.

Qian, Yue, and Zhenchao Qian. 2017. Assortative mating by education and Hukou in Shanghai. *Chinese Sociological Review* 49 (3): 239–262. https://doi.org/10.1080/21620555.2017.1288066.

Qian, Zhenchao, Daniel T. Lichter, and Leanna M. Mellott. 2005. Out-of-wedlock childbearing, marital prospects and mate selection. *Social Forces* 84 (1): 473–491. https://doi.org/10.1353/sof.2005.0117.

Rauscher, Emily. 2015. Effects of Early U.S. Compulsory schooling laws on educational assortative mating: The importance of context. *Demography* 52 (4): 1219–1242. https://doi.org/10.1007/s13524-015-0402-5.

Rauscher, Emily. 2020. Why who marries whom matters: Effects of educational assortative mating on infant health in the United States 1969–1994. *Social Forces* 98 (3): 1143–1173. https://doi.org/10.1093/sf/soz051.

Ravazzini, Laura, Ursina Kuhn, and Christian Suter. 2017. Do opposites attract? Educational assortative mating and dynamics of wage homogamy in Switzerland, 1992–2014. *Swiss Journal of Sociology* 43 (3): 567–586. https://doi.org/10.1515/sjs-2017-0028.

Raymo, James M., and Miho Iwasawa. 2016. *Diverging destinies: The Japanese case*. Springer.

Raymo, James M., and Yu Xie. 2000. Temporal and regional variation in the strength of educational homogamy. *American Sociological Review* 65 (5): 773–781.

Schwartz, Christine R. 2010. Earnings inequality and the changing association between spouses' earnings. *American Journal of Sociology* 115 (5): 1524–1557. https://doi.org/10.1086/651373

Schwartz, Christine R. 2013. Trends and variation in assortative mating: Causes and consequences. *Annual Review of Sociology* 39 (1): 451–470. https://doi.org/10.1146/annurev-soc-071312-145544.

Schwartz, Christine R., and Hongyun Han. 2014. The reversal of the gender gap in education and trends in marital dissolution. *American Sociological Review* 79 (4): 605–629. https://doi.org/10.1177/0003122414539682.

Schwartz, Christine R., and Robert D. Mare. 2005. Trends in educational assortative marriage from 1940 to 2003. *Demography* 42 (4): 621–646.

Senda, Yukiko. 2002. Influence of childcare resources on the employment continuity of married women: Focusing on occupation and birth cohort. *Journal of Population Problems (Jinko Mondai Kenkyu)* 58 (2): 2–21 (in Japanese).

Shafer, Kevin. 2013. Unique matching patterns in remarriage: Educational assortative mating among divorced men and women. *Journal of Family Issues* 34 (11): 1500–1535. https://doi.org/10.1177/0192513X12459020.

Shafer, Kevin, and Spencer L. James. 2013. Gender and socioeconomic status differences in first and second marriage formation: Comparing first and second marriage formation. *Journal of Marriage and Family* 75 (3): 544–564. https://doi.org/10.1111/jomf.12024.

Shirahase, Sawako. 2014. *Social inequality in Japan*. London: Routledge.

Smits, Jeroen, and Hyunjoon Park. 2009. Five decades of educational assortative mating in 10 East Asian societies. *Social Forces* 88 (1): 227–255.

Tachibanaki, Toshiaki. 2005. *Confronting income inequality in Japan*. New York: MIT Press.

Tachibanaki, Toshiaki, and Sayaka Sakoda. 2013. *Fufu Kakusa Shakai (Couples Disparity Society)*. Chuo Koronsha (in Japanese).

Tsutsui, Junya. 2016. *Kekkon to Kazoku No Korekawa (The Future of Marriage and Family)*. Kobunsha. (in Japanese).

Uchikoshi, Fumiya. 2019. Consequences of educational assortative mating through divorce. *Japanese Sociological Review* 70 (1): 10–26 (in Japanese). https://doi.org/10.4057/jsr.70.10.

Uchikoshi, Fumiya. 2020. Explaining declining educational homogamy: The role of institutional changes in higher education in Japan. *SocArxiv* 45. https://doi.org/10.31235/osf.io/45axy.

van Bavel, Jan, Christine R. Schwartz, and Albert Esteve. 2018. The reversal of the gender gap in education and its consequences for family life. *Annual Review of Sociology* 44(1):341–60. https://doi.org/10.1146/annurev-soc-073117-041215.

Van Zantvliet, Pascale I., Matthijs Kalmijn, and Ellen Verbakel. 2014. Parental involvement in partner choice: The case of Turks and Moroccans in the Netherlands. *European Sociological Review* 30 (3): 387–398. https://doi.org/10.1093/esr/jcu045.

Wong, Raymond Sin-Kwok. 2003. To see or not to see: another look at research on temporal trends and cross-national differences in educational homogamy. *Taiwanese Journal of Sociology* 31: 45–88.

Xie, Yu. 2000. Demography: Past, present, and future. *Journal of the American Statistical Association* 95: 670–673.

Yoshida, Takashi. 2011. Trend of educational and occupational assortative mating in Japan. In *Studies of social stratification in contemporary Japan: from the perspective of micro-macro linkages vol. 3, Education, Gender, and Marriage*, ed. Y. Sato, 263–76. The Research Group of Studies of Social Stratification in contemporary Japan.

Appendix

In estimating the odds of educational assortative mating and their change over time, we fit nine models (model fit statistics corresponding to each model can be seen in Table 5.3). Below is a description of each model.

Model 1: Conditional independence

$$\ln F_{ijk} = \lambda + \lambda_i^H + \lambda_j^W + \lambda_k^C + \lambda_{ik}^{HC} + \lambda_{jk}^{WC}$$

Model 1 is called the conditional independence model, which adjusts husbands' and wives' educational distribution $(\lambda_i^H,\ \lambda_j^W)$, marriage cohort, (λ_k^C) , and its interaction with each spouse's educational attainment $(\lambda_{ik}^{HC}, \lambda_{jk}^{WC})$. We build upon this basic model to examine whether different specifications of educational pairing patterns (and their change across marriage cohorts) improve model fit. Model 2 adds homogamy parameters (δ_{ij}^{HW}) , which allow the odds of homogamy to vary across levels of educational attainment, thus using four degrees of freedom. In Model 3, we examine whether homogamy captured by this design matrix has changed over time by including interactions with marriage cohort (δ_{ijk}^{HWC}) .

Model 2: Homogamy

$$\ln F_{ijk} = \lambda + \lambda_i^H + \lambda_j^W + \lambda_k^C + \lambda_{ik}^{HC} + \lambda_{jk}^{WC} + \delta_{ij}^{HW}, \text{ where } \delta_{ij}^{HW} = 0 \text{ for } i \neq j$$

Model 3: Cohort change in homogamy (a)

$$\ln F_{ijk} = \lambda + \lambda_i^H + \lambda_j^W + \lambda_k^C + \lambda_{ik}^{HC} + \lambda_{jk}^{WC} + \lambda_{ij}^{HW} + \delta_{ijk}^{HWC},$$
$$\text{where } \delta_{ijk}^{HWC} = 0 \text{ for } i \neq j$$

The next three models examine different patterns of change in assortative mating across marriage cohorts. Model 4 constrains the pattern of cohort change in homogamy to be the same for all levels of educational attainment. Models 5 and 6

F. Uchikoshi and J. M. Raymo, *Educational Assortative Mating in Japan*,
Population Studies of Japan, https://doi.org/10.1007/978-981-16-3713-1

examine change across marriage cohorts in educational hypergamyand hypogamy, respectively (while constraining homogamy to be constant across cohorts).

Model 4: Cohort change in homogamy (b)

$$\ln F_{ijk} = \lambda + \lambda_i^H + \lambda_j^W + \lambda_k^C + \lambda_{ik}^{HC} + \lambda_{jk}^{WC} + \lambda_{ij}^{HW} + \delta_{ijk}^{HWC},$$
$$\text{where } \delta_{ijk}^{HWC} = 0 \text{ for i} \neq \text{j, 1 for i} = \text{j}$$

Model 5: Changing hypergamy

$$\ln F_{ijk} = \lambda + \lambda_i^H + \lambda_j^W + \lambda_k^C + \lambda_{ik}^{HC} + \lambda_{jk}^{WC} + \lambda_{ij}^{HW} + \delta_{ijk}^{HWC},$$
$$\text{where } \delta_{ijk}^{HWC} = 0 \text{ for i} \leq \text{j, 1 for i} > \text{j}$$

Model 6: Changing hypogamy

$$\ln F_{ijk} = \lambda + \lambda_i^H + \lambda_j^W + \lambda_k^C + \lambda_{ik}^{HC} + \lambda_{jk}^{WC} + \lambda_{ij}^{HW} + \delta_{ijk}^{HWC},$$
$$\text{where } \delta_{ijk}^{HWC} = 0 \text{ for i} \geq \text{j, 1 for i} < \text{j}$$

Model 7 is called the crossing model, which often performs better than other models in terms of predicting educational pairing frequencies (Schwartz and Mare 2005). This model explicitly models the permeability of barriers to marriage across educational levels. Model 8 examines whether crossing parameters have changed across marriage cohorts.

Model 7: Crossing model

$$\ln F_{ijk} = \lambda + \lambda_i^H + \lambda_j^W + \lambda_k^C + \lambda_{ik}^{HC} + \lambda_{jk}^{WC} + \gamma_{ij}^{HW},$$

where $\gamma_{ij}^{HW} = \sum_{q=j}^{i-1} \gamma_q$ for $i > j$, $\gamma_{ij}^{HW} = \sum_{q=i}^{j-1} \gamma_q$ for $i < j$, and $\gamma_{ij}^{HW} = 0$ for $i = j$.

Model 8: Cohort change in crossing

$$\ln F_{ijk} = \lambda + \lambda_i^H + \lambda_j^W + \lambda_k^C + \lambda_{ik}^{HC} + \lambda_{jk}^{WC} + \gamma_{ijk}^{HWC},$$

where $\gamma_{ijk}^{HWC} = \sum_{q=j}^{i-1} \gamma_q$ for $i > j$, $\gamma_{ijk}^{HWC} = \sum_{q=i}^{j-1} \gamma_q$ for $i < j$, and $\gamma_{ijk}^{HWC} = 0$ for $i = j$.

Finally, we apply the log-multiplicative layer effects models (Xie 1992), which allow us to estimate changes in the strength of the associations over time while assuming that the overall patterns of educational pairing are constant. This model is attractive for its parsimonious representation of cohort change in pairing patterns and is expressed as follows:

Model 9: Log-multiplicative association

$$\ln F_{ijk} = \lambda + \lambda_i^H + \lambda_j^W + \lambda_k^C + \lambda_{ik}^{HC} + \lambda_{jk}^{WC} + \beta_k^C \lambda_{ij}^{HW}$$

Here, λ_{ij}^{HW} are parameters for educational homogamy as in Model 2 and β_k^C is the log-multiplicative parameter. This model produces a single-parameter representation of cohort differences in the strength of the educational homogamy (Powers and Xie 2008; Xie 1992). The β parameter is set to 1, with the earliest marriage cohort as the reference, and we evaluate changes in the association in terms of the percent change in this parameter relative to the value for the reference cohort.

References

Powers, Daniel A., and Yu Xie. 2008. *Statistical methods for categorical data analysis*. 2nd Edition. Emerald Group Publishing.
Schwartz, Christine R., and Robert D. Mare. 2005. Trends in educational assortative marriage from 1940 to 2003. *Demography* 42 (4): 621–646.
Xie, Yu. 1992. The log-multiplicative layer effect model for comparing mobility tables. *American Sociological Review* 57 (3): 380–395.

F. Uchikoshi and J. M. Raymo, *Educational Assortative Mating in Japan*,
Population Studies of Japan, https://doi.org/10.1007/978-981-16-3713-1

Index

F. Uchikoshi and J. M. Raymo, *Educational Assortative Mating in Japan*, Population Studies of Japan, https://doi.org/10.1007/978-981-16-3713-1

The manufacturer's authorised representative in the EU is Springer Nature Customer Service Centre GmbH, Europaplatz 3, 69115 Heidelberg, Germany. If you have any concerns regarding our products, please contact ProductSafety@springernature.com

Printed and bound by CPI Group (UK) Ltd, Croydon, CR0 4YY
30/09/2026
02225261-0012